GOLD BARS

≈

CHARTERING YOUR BOAT

FOR MONEY

Captain Conrad N. Brown, Jr.

SP

Shipyard Press

www.shipyardpress.com

GOLD BARS

CHARTERING YOUR BOAT FOR MONEY

Published by:
Shipyard Press
www.shipyardpress.com

10 9 8 7 6 5 4 3 2

Printed in the United States of America

ISBN Print Edition: 978-0-9769903-1-4

Library of Congress Cataloging-in-Publication Data
Brown Jr., Conrad N.
GOLD BARS / Conrad N. Brown, Jr.
1. Charter boat business-United States. 2. US Coast Guard
License-United States. I. Title.

For quantity discounts on bulk purchases of this book please
contact the publisher at www.shipyardpress.com

Table of Contents

Disclaimer

This book was written to provide information about the business of chartering boats. It is sold with the understanding that the author and the publisher are not engaged in the profession of rendering services in the legal, accounting, tax preparation or other professional services. If you have questions relating to those areas, contact your own attorney, accountant or other professional.

It is not the purpose of this book to be a complete manual for starting and operating a charter or other business. The information presented is intended to stimulate thinking about issues that might be addressed and suggest solutions that could be applied. In particular, a discussion regarding obtaining a Coast Guard license to legally operate boats is intended as advice and direction on how to access official Coast Guard information on the subject.

It is the purpose of this book to educate and entertain. The author and the publisher have neither liability nor responsibility to any person or entity with regard to how this information is used nor are they responsible for any loss or damage caused or alleged to be caused directly or indirectly by any information contained in the book.

If you do not wish to be bound by the above, you may return this book to the publisher for a full refund.

Acknowledgements

Without the help of many, many people, books like this could not be written. I would like to thank, in the Virgin Islands, Capt. Bob and Joanie Jill who were my initial role models along with Captain's Fergus Walker, Poseidon and Ross Norgrove, White Squall, and Kathy, my mentor and mate at the critical start-up time, kick starting my dream. I would also like to thank the many charter guests we had onboard, I learned from each one valuable life lessons.

Captain and author Dave Ferneding and his wife Roz let me get up close to big time chartering on their yacht Antares, and as author-advisor, Dave has had a big hand in this book as well.

My two sons, Conrad and David, both captains earned their licenses on their own and have sailed much farther than their Dad ever did. They are everywhere in this book. Researching, I also relied on good help from Bill Chubb, Lt. Mike Washburn, Ens. Patrick Drayer, and CWO Sally Gudgel of the Coast Guard. Capt. Bob Arnold of the St. Petersburg Sea School gave valuable assistance along with Jim Araiza of SailAway Yacht Charters.

In Florida, I was helped by Harbormasters Lief Lustig and Jim Langer and all of the crew at the City Yacht Basin in Ft. Myers. Tony Mantia showed up one day with a loaded tool belt providing much needed shipwright relief just when the restoration project was growing exponentially.

Finally, to my wife Marilyn, my last mate for her support in my many schemes, thank you sweetheart, and thank you reader, for picking up this book. I hope it is of some help with your dreams.

Other books by the author

Maritime Stories

My Wright Stuff

SP

Shipyard Press

www.shipyardpress.com

Dedicated to my Dad,

and to all the dads who have given

their sons and daughters

a love for the sea.

Chapter 1

Getting Started

"Where the pupil is willing, the teachers will appear."

Italian proverb

Imagine

Can you see yourself as a yacht captain? You are wearing your crisp white shirt with four gold stripes on your epaulets directing the docking of a million dollar yacht or perhaps you see yourself dressed in sneakers and T-shirt out fishing with newfound friends—for money.

Or you can envision yourself and your lovely mate sailing tropical seas, calling at exotic ports, entertaining guests with salty sea stories while on the stereo Jimmy Buffet sings about the life you've chosen.

Maybe your dream is taking people up a lazy river to experience the natural, unspoiled beauty that still exists on some hidden waterway. Or all you've ever wanted to do was have your own sport fisherman and go out for the big ones with some paying guests to make it all possible.

You can earn a living by chartering a boat, yours or another's. My dream of becoming a charter boat captain took ten years to come true, but when I was finally aboard my boat in the Virgin Islands with a newly minted and framed license on the bulkhead, I was the happiest man alive.

Your dream may take more or less time but if you use this book as a tool, persist, and hold onto your dream, you will get there. I know where the money is in running a charter boat business and will show you how to make enough to support your chosen lifestyle.

Fishing Charters

Imagine you are up before dawn and down at your boat, the Fish'n Dream. The air is crisp from a cold front that came through during the night and it would feel even colder if you weren't moving around briskly checking your boat and its fishing gear.

Your mate shows up so you send him over to the bait shop for a tub of pompano while you finish checking your boat, preparing for the day ahead.

Everything checks out OK as you run through your mental list; engine and tranny oil, fuel in the tanks, heads working, drinks and sandwiches in the cooler.

From the fly bridge, you start and warm up both engines, then leaving them idling, run checks on all electronic systems to make sure they are good to go. Then it's time to go down below to make a pot of coffee.

The mate hops back aboard, stows his bait and squares away the cockpit for fishing. You see a bunch of guys, sleepily stumbling down the dock in the near darkness; they must be your fishing party. Stepping back aft, you wave them over. Time to go to work.

After introductions all around you write each name on a piece of duct tape and stick it on their shirt back. There are some rude jokes about the nametape but they are soon forgotten, as we get ready to get underway. Some of the crowd this morning seems to know their way around a fishing boat, which will make life easier. Others are neophytes and will probably need a lot of help. The main goal is to make sure everyone has a good time and catches a fish or two.

Everybody gets a cup of coffee then the mate slips the lines; you back her out and head for the inlet that leads out to the ocean and the Gulf Stream. Another day in paradise begins as that big beautiful sun breaks the horizon over Bimini, 55 miles away.

Fish'n Dream cruises at 18 knots fully loaded and since the Gulf Stream is about twelve miles offshore today, you keep the speed up while the mate is rigging gear and baiting up. We want to get hooks in the water early. The sea breeze this morning is northeasterly, about 3-5 knots now but you know it will build as the sun heats the air over the land.

The sea around you is relatively flat, with just a small surface chop and the hint of a swell from the southeast.

Your anglers are getting used to the boat and working on their sea legs. Some join you on the bridge for snatches of conversation as the wind whips speech into broken words like a fading cell phone.

Forty minutes later the water color changes to that incredible deep indigo blue and the GPS indicates we have a northerly drift of 1.2 knots courtesy of the Gulf Stream. Setting a course directly into the wind and across the stream, you slow the boat down to begin the first troll of the day and, well, you know the rest of the story.

After a great day at sea, having caught enough fish and taken enough pictures to prove their stories, you are back at the dock at 1600 hours with a tired but happy cockpit full of smiling sunburned anglers who have gladly paid $300 apiece for the experience. Plus a couple hundred for your tip and a hundred more for the mate. As they walk away, down the dock, you chuckle at their names still taped to their shirts, walking advertisements for your business. Friends will ask about the tape and they get to tell of their day on the water....on your boat.

Your gross for the day is $300 x 6 = $1,800. The costs are; $240 fuel, $200 for the mate, $50 for bait and snacks, $100 to cover overhead (insurance, dockage, advertising, etc.)

This days work leaves you with $1,210 for the charter boat business, $200 cash in your pocket (remember its income and taxable) and a nice fresh mahi-mahi in the cooler for your grill tonight. Two of the six will be back next year and bring their friends. Then you can do it all over again. Is this fun or what?

Sailing Charters

You and your mate have made the long thirteen hundred mile rhumb line passage down to the Virgin Islands.

It was much colder and rougher than you expected, but things eased, as you got further south and closed with the Tropic of Cancer. The skies became bluer, the clouds whiter, the water warmer, and that lazy, liquid feeling of the tropics gradually washed away the rough cold weather memories. A day away from landfall you detect the sweet, musty smell of land over the horizon.

Now you've spent the last few weeks resting, meeting people in the yachting community, and preparing your boat for the St. Thomas Charter Boat Show held in November each year. You meet a ton of brokers and agents, some of which like you and your boat. They give you lots of encouragement and one even books your boat for a Thanksgiving charter this season.

Your First Charter

After the show, you leave Charlotte Amalie on your very first charter with two couples anxious to have a good time on their dream sailing vacation. With the water, the sun and a thousand beaches and reefs to explore ahead, plus sailing your own yacht in paradise, you know you've made the right choice to be here.

Leaving the harbor and Red Nun buoy #2 to port marking Triangle Reef, you lay your course also passing Buck Island to port while sailing close-hauled for French Cap Cay.

Rounding French Cap up close to see the birds and nests built into the crags and rocky slopes, you slack the sheets for a great broad reach back to Christmas Cove at St. James Island. Spray across the foredeck keeps the guests in the cockpit and the lee rail buries in the foam every time the boat surges forward on another great draft of the trade winds. Your yacht climbs out of the troughs and flies across the crest of swells coming five thousand miles to dance for your pleasure. You laugh with the guests and guess as to who is having the most fun.

They are having the time of their lives but are just a little scared also, adding to their thrill. From French Cap it's a great run back to Christmas Cove.

We drop a hook in the southern bight of the Cove next to the big rock, a great place for their first snorkel and swim of the trip. You and the guests snorkel around it looking for the elusive giant manta ray who sometimes calls it home. Afterward your guests, who are fast becoming your good friends, are ready for a drink on the aft deck watching the sunset.

Down below your mate is singing to herself, as she preps for dinner. Everyone is exhausted and sleeps soundly after a terrific dinner and an evening of exchanging interesting life stories. You drift off dreaming of the prospects of six more lovely days, clearing into and circumnavigating the British Virgin Islands.

The next day it's out through the Cut into Pillsbury Sound a small inland sea ringed with beaches.

We make a short tack then into St. John, for a quick clear out of US customs and last minute shopping and mailing post card for the guests.

Hauling out of St. John it up past Caneel Bay resort with patrons and presidents lolling on the beach wishing there were aboard that beautiful sail boat (yours) just offshore.

Six more days of this and by the last day, you've made some great new friends, and they have hundreds of fresh new memories plus a slew of pictures to download and email their friends when they get home. You have another four thousand dollars in the kitty after expenses plus a four hundred dollar tip to spend celebrating their departure.

This was a great first charter and while you were glad to see your charterers come aboard, you are even happier to see them go, getting your boat back and your new friends out of your personal space.

Sometimes, if they have time, after a charter many couples hurry off to a private cove or anchorage to clean the yacht thoroughly, removing all traces of the charterers.

Then, following a swim in clean, clear water and a shower to wash away all traces of work, they make love in perfect privacy for the first time in a week. It is a process of clearing the mind, much as clearing the palate during a long meal.

Other Charters

Not everyone has the same dream and yours may be flats fishing in the Florida Keys or trolling the   Bahama bank, taking people on sightseeing cruises around historic ports or up interesting rivers, or just cruising from port to port.

In some parts of the country, there are whale-watching trips, leaf peeper trips in the fall, and charters to sail in or regattas to watch aboard a charter yacht.

Some captains find working on classic sailing yachts to their liking. In Newport Rhode Island, a group of twelve-meter, ex-Americas Cup yachts duke it out every day during season with twelve lucky passengers-crew aboard, and a charter captain grinning from ear to ear.

Then there are adventurers who sail racing yachts all over the world and chase the silver trophies. There are hundreds of opportunities to use you license "to carry passengers for hire," as the Coast Guard says. Perhaps you will find a corporate owned yacht and draw an impressive salary for always having the boat ship-shape and ready to go anywhere at a moments notice. Then again you just may have the next great charter idea up your sleeve and reading this book will encourage you to make it happen. Good luck.

Other Boating Jobs

Other charter jobs include delivering sail or powerboats from port to port or halfway around the world. Every new boat is delivered somewhere, if not by the owner then by a captain hired to do the job. Bareboat charter companies often move some of their fleets to follow the sailing seasons. There are opportunities everywhere if you are ready.

You can get one of these jobs or create a business of your own by working up a reasonable plan and following it for your chartering success. The book you are holding is a model for a successful charter business.

This book will also show you how to earn your Coast Guard license if you do not already have one. The license is necessary. Get the Coast Guard application going first while you tackle the rest of the business.

Ready? Let's get started. I'll help you

Chapter 2

Coast Guard License

"Sailors with their built-in sense of order, service and discipline, should really be running the world." Nicholas Monsarrat

What License is Required

To operate a private yacht (your own) or to captain for an owner and non-paying guests, technically you do not need a license. In practice however the boat owner's insurance company will, in most cases, insist upon the captain holding the appropriate Coast Guard license as a condition for granting the owner an insurance policy for the boat operated by a paid captain.

Before you can start chartering for money on your own boat in the United States including Alaska, Hawaii, the Virgin Islands, Commonwealth of Puerto Rico and Pacific Territories you must first obtain a license from the U.S. Coast Guard.

In other words, to charge people to ride on your boat, no matter how large or small, in the U.S., you must have a U.S. Merchant Marine Officer license commonly called a Captain's license, six-pack, or an Operators license.

You may apply for the license at any U.S. Coast Guard Regional Examining Center (REC). Start at this USCG website:

http://www.uscg.mil/nmc/cb_capt.asp.

Master Near Coastal

Issued for 25, 50 or 100 gross tons

The Master Hundred Ton license is for both inspected and un-inspected vessels. Near coastal means, you are licensed to take passengers offshore within 200 miles of the coastline. Basic requirements for this license are:

- 720 days working onboard total time
- 90 of those days in the last three years
- 360 days in near coastal waters (offshore)
- U.S. Citizenship is required
- Minimum age to apply is 19
- A written exam is required

The license is called a "Masters License" because the wording on the license says:

"This is to certify that ***your name*** having been duly examined and found competent by undersigned, is licensed to serve as:

MASTER OF NEAR COASTAL STEAM OR MOTOR VESSELS OF NOT MORE THAN –100-GROSS TONS."

Other useful information on the license is the serial number in the upper left hand corner, the date of issue, the expiration date and the issue number in the upper right hand corner. The license is good for five years, so each issue covers a five-year period. A (-2-) in the corner means it is the second five-year term of the license.

The endorsements will be noted on the license. Sailing and towing endorsements or other restrictions are found on the front immediately following the word MASTER.

(As of April 15, 2010 Homeland Security, of which the Coast Guard is a part, will issue licenses in a form similar to U.S. Passports and all endorsements will be shown therein.)

With a 100-ton masters license you could theoretically be master of an inspected passenger vessel carrying a large number of passengers so long as it fell within the 100 tons as measured by the Coast Guard and the vessel's inspection certificate.

In practice, all vessels that are "inspected" by the Coast Guard will have a maximum capacity for both passengers and crew posted on the certificate for the vessel and this is your passenger limit for that vessel.

Many of the small passenger cruise vessels used on rivers and calling at coastal ports are especially built to fall within this 100-ton category.

With your 100-ton license, you could be captain of a small cruise ship plying the waters of the Gulf of Maine, Chesapeake Bay, many of our inland rivers or the Inland Passage to Alaska.

If your boating experience is in smaller vessels exclusively, you may be issued a license for 25 or 50 tons. At your next license renewal, you can ask for your tonnage to be raised.

Master Inland

Also called a "Masters License."

This license is essentially the same as a hundred ton license except for restrictions noted below. Basic requirements follow:

- For inspected or uninspected vessels inland 360 days total time required
- 90 days within the last three years
- U.S. Citizenship required
- Minimum age is 19 to apply
- A written exam is required
- Endorsements: OUPV, Mate Near Coastal

OUVP Near Coastal (6-pack)

For uninspected 6 passenger vessels, inland and offshore up to 100 miles.

- 360 days total time
- 90 within the last three years
- 90 days must be Near Coastal
- (offshore up to 100 miles)
- U.S. Citizenship not required
- Minimum age is 18 to apply
- A written exam is required

OUVP Inland (6-pack)

For uninspected six passenger vessels on inland waters.

- 360 days total time boating experience
- 90 days within the last three years
- Offshore time not required
- U.S. Citizenship not required
- Minimum age to apply is 18
- A written exam is required

Launch Operator OUVP Limited

Uninspected launches six passengers or less.
* 90 days experience in a boat
* Minimum age to apply is 17
* A written exam is required

Launch Operator-Master Inland Limited

Inspected and Uninspected launches
* 120 days experience in a boat
* Minimum age to apply is 18
* A written exam is required

OUVP means:

Operator Un-inspected Vessel Passenger

Inspected Vessels

What is an inspected vessel? It is not the "safety inspection" offered by the Coast Guard Auxiliary or other boating group.

Inspected means a vessel the Coast Guard has physically examined and has established the maximum number of passengers that may be carried and the waters in which it may operate, carrying passengers for hire. The inspection process usually takes place at the time of building of the vessel and Coast Guard inspectors monitor each major step of the construction.

If the vessel meets all their safety requirements, they will issue an inspection certificate for the vessel, which certifies it for a certain number of passengers and the area of operation, (inland, coastal waters, offshore). There are other considerations required for inspected vessels such as; collision bulkheads, fire fighting equipment, and other safety requirements, which are outside the scope of this book.

It may be expensive to change your uninspected vessel, to meet the requirements for an inspected vessel, it's time consuming, and not at all a sure-fire process.

For more information on this contact your nearest Coast Guard Marine Safety Office (MSO). (Also listed in the appendix) One general rule to remember is; to carry over six passengers for hire on a vessel, generally it must be an "inspected" vessel.

The Coast Guard issues many types of licenses other than those above, merchant marine shipping licenses, engineering licenses, towing licenses, radio operator, fire fighting, radar observer or mates and able seaman licenses. These apply to commercial maritime shipping.

This book pertains to pleasure charter boats and related issues. However, when you earn a Coast Guard Master license you will become a:

"United States Merchant Marine Officer"

As a United States Merchant Marine Officer, theoretically you could be called up to serve in time of war. When you submit your application, you will be given the option of participating in the Mariners Tracking System. In the event of a national emergency, you might be offered a job in the Maritime Administration (MARAD). Wouldn't that be neat! This is one reason the requirements for the license are so comprehensive and another reason for the U.S. citizenship requirement. When you earn a U.S. Coast Guard License, you are a member of an elite fraternity. Be proud of your accomplishment.

How to Obtain Your License

Start by asking for an information package from one of the Coast Guard Regional Examination Centers (REC) at http://www.uscg.mil/nmc/cb_capt.asp. For your convenience, use a REC nearest your home or where you intend to work. Rules are constantly changing and early in 2010 you are required to obtain Merchant Marine Credentials (MMC) before you can apply to the Coast Guard for a license. Start at:

http://www.TSA.gov

After you get your MMC credentials, when you contact the REC, they will start a file for your application. <u>Continue to use the same REC and the same name for all your questions and submissions.</u>

Switching from one to another will unnecessarily complicate your application and may result in something being lost. It has happened!

Fill out the application carefully. If you miss a section or send it in incomplete, everything stops while the application is returned to you for correction.

Requirements Common to All Licenses

Like everything, rules and requirements change from time to time, so log on to the Coast Guard web site at:

http://www.uscg.mil/nmccb_capt.asp

There you will find the latest requirements and downloadable Coast Guard application forms.

The next pages tell you how to fill out the Application. There are also tips from the CG on common mistakes made on their application. Read and follow their suggestions.

Section I

Use your full legal name. Hold nothing back in any questions you are asked.

Section II

Example: Master 100 ton near coastal or inland waters, sail, steam or motor vessel. You may not get everything you ask for, but the Coast Guard will not, on their own give you more than what you ask for, so apply for all that you feel qualified.

Section III

You must have no narcotic drug addiction, and must present proof of being cured if you ever used or were addicted to a narcotic drug.

No narcotics convictions within the last three to ten years depending on the conviction. Form CG-719P will give you details of what is required and what is accepted. <u>You will be tested for drug use. If you use illegal drugs, now is a good time to stop.</u> If you don't you will not get a license and frankly I don't want you "high,"out on my water!

Criminal Convictions

Other than minor traffic violations, criminal convictions must be shown including DUI, expunged records, pre-trial diversions, etc.

You must include a written statement listing every conviction including the crime, name and location of the court, date of conviction, sentence received, the date you were released from jail and a narrative relating to what occurred the day of the crime. If in doubt, include it.

You must make a written statement of any suspension or revocation of your driver's license or refusal to submit to an alcohol or drug test outlining where and when each event occurred and a brief account of what happened. Submit a copy of your state driver's license or a letter stating your eligibility to possess a driver's license.

Traffic Convictions

A written statement is required from fatal accidents, reckless driving, racing on a highway, operating a motor vehicle under the influence, or impaired by alcohol or a controlled substance.

If you fail to report any of the above items that apply to you on your application and the reviewers later determine you have been less than truthful, then your entire application is suspect and may result in your application being denied and could result in you being barred from ever being able to qualify for a license.

The Coast Guard REC will evaluate all your information before granting you a license. Bad things happen to good people, but if you are upfront and show evidence of responsibility, you will probably sail right through. A Coast Guard licensed Master is a responsible person, you wouldn't want it any other way.

Section IV

Character References

You must obtain three character references from persons who have knowledge of your suitability to serve as a Master. If your qualifying time is in a commercial capacity, your references should be from licensed officers, at least one of which should be a Master.

Section V

Mariners Consent

You will be fingerprinted by the REC when you appear in person to verify your identity.

Your prints will be sent to the FBI and will be entered in the National Fingerprint Database.

National Drivers Register

The Coast Guard will check your application against the National Drivers Register and other criminal record databases for serious crimes and some motor vehicle offences. If you think you may be flagged for some driving offense, check with your REC or captains school for direction.

Proof of U.S. Citizenship

You must be a U.S. citizen, born or naturalized to hold a Masters Inland or Near Coastal license. Other 6-pack and launch operators are not required to be U.S. Citizens.

English Spoken Here

You must be able to read and speak English. The exception is Puerto Rico where Spanish is accepted for 6-pack licenses only, but those licenses are restricted to Puerto Rico and surrounding waters.

Physical Exam

A licensed M.D., D.O., or physician's assistant under a physician's direction must certify that you are of sound health. <u>Have your doctor sign the form and put his/her license number on the form</u>. You must be physically fit to perform your duties and must be physically and mentally able to stay alert for 4 to 6 hour shifts and be free of any medical condition that would pose a risk of sudden incapacitation to perform your duties, (seizure, heart attack, stroke or coma).

A copy of the medical form is in the appendix. Review it with your physician before the examination to be sure all the required tests are available.

If you have any questions regarding a pass/fail test, contact your REC before having an examination.

You must have correctable vision to at least 20/40 in each eye and no more than 20/200 uncorrected. Wavers to 20/800 uncorrected are available. Your color sense is important also but even color blindness can be overcome with an OUVP license restricted to daytime only.

Your physical exam must state that you are "competent" and that your color vision test or uncorrected vision tests are in accordance with the requirements.

The medical exam will address vision, hearing, prescription medications, asthma, diabetes, high blood pressure, pulse rate/rhythm, and any type of heart or vascular disease. Find a physician who is familiar with the Coast Guard requirements or the application may be sent back for some minor correction, delaying your application several weeks or months.

Other bars to a license are very poor hearing, epilepsy, infectious disease, drug abuse, alcohol abuse or other psychiatric disorders. You must be physically able to perform your duties beyond the usual. The safety of your passengers and the vessel may well depend on your physical abilities in time of crisis.

If some physical or mental deficiency exists, you may appeal to the Commandant of the Coast Guard for an opinion. Try this only after you have convinced the REC officer in charge that extenuating circumstances exist that will not endanger the public.(Underlined sections above are common reasons physical exams are returned for correction, make sure you have all these items covered with your doctor.)

First Aid

You must have completed a first aid course within the past 12 months from the American Red Cross or a Coast Guard approved school. You must possess a valid CPR certificate for successful course completion from the American Red Cross or other approved school.

Written Examination

The Coast Guard requires that you pass a written exam. For the past several years the Coast Guard has approved schools, commonly called "Captain's Schools or Sea Schools®" to administer the written examination for the Coast Guard. Many schools offer day or night classes, which grant a diploma at the end of the class if you pass the final exam. The diploma is not a Coast Guard license. Most schools will assist you in completing and submitting your application.

You must supply two forms of identification, one of which must be a government photo ID. (Drivers license) You are not required to furnish separate photographs for any of these four licenses.

You must prove your U.S. citizenship with a birth certificate, original passport or baptismal certificate. The information packet from the TSA, Transportation Safety Administration, **www.tsa.gov** will give you're the details.

Social Security Card

A social security card must be submitted. If you have lost yours, contact your nearest S.S. office for a replacement.

Time on the Water

All licenses have a requirement for actual sea time onboard a vessel to make sure you have some idea of what you are doing and how to apply any book learning you may have. The amount of time varies with the license. A lifetime accumulation of time will help you qualify.

However, at least ninety days of your qualifying time must be acquired within the past three years prior to your application. To some degree, each REC sets their own standards for sea service in their area, check with them if you have questions.

You can fulfill your Sea Service time on your own vessel if it is five tons or greater. Documenting your own time on the water is necessary or you may have someone familiar with what you do and where you go who can certify your time. Your Dockmaster for instance. Time spent in the Navy, the Coast Guard or other military or sea service will count as long as you can document the time your vessel was underway. A letter from your C.O. will usually suffice.

The Coast Guard just wants to know you have actually served aboard a boat, a real boat, preferably operated by someone who has a license and knows what they are doing. It's like learning to drive, you go out with someone who knows how and can pass on that information to you as you practice.

A day of service is normally eight hours, a twelve-hour shift or watch is a day and a half. All REC's will accept four hours as a day on vessels up to 100 tons. Check with your REC. Make sure the dates, tonnages, and other facts agree with documents in the submission package.

Reviewers are looking for discrepancies and errors that may indicate illegal activities or less than full disclosure. Remember you are dealing with government employees here. Cross your T's, dot your i's and have someone check your work before you send it in.

Most REC's give you a phone number to call and check on your application. Call that number occasionally to make sure your application is moving through the system and not hung up on some minor point.

After submitting the application, you must pass the exam or furnish a diploma showing successful course completion. The exam and any captain's course you take will consist of questions regarding navigation and chart navigation, general deck questions, environmental questions, rules of the road and safety. If you are going for an auxiliary sail endorsement or towing assistance endorsement, you will have additional questions on your exam. If you are applying for a Great Lakes and Inland rivers license, there will be specific questions regarding those areas.

The Coast Guard conducts examinations or will accept a passing grade from an approved school. The best way to be assured you will pass the test is to take a course from a captain's school near you. Most schools have a no pass, no pay policy, so you may re-take the course and exam if you fail the first time.

After passing your written test, submit your diploma from a captains school and comply with every other CG requirement; proof of citizenship, meet the minimum age, document your time served on the water, pass the drug test, get fingerprinted, pass the physical exam, earn your CPR & first aid certificates, and pay the fees. Then wait a few months and finally receive the license you qualify for, the OUVP, Limited Master or 100-ton license.

Remember, just passing the written course at a captains school is only the start of your journey. You must follow through all the procedures to earn your license.

New Orleans and Miami are two of the Regional Examining Centers with the most activity. If you have a choice by virtue of where you live or plan to work and can chose a center other than either of those, your processing time will be shorter. New Orleans receives some 15,000 applications a year while Miami at least 5,000.

Other REC's may have more time and more personnel to handle your application quickly. As tough as it may seem sometimes, licenses are possible to get. The Coast Guard awards approximately 7,200 new Masters or OUPV licenses each year.

Each year the standards are raised. Now is a better time than "sometime" to get your license.

Remember the answer to any question can be found at:

http://www.uscg.mil/nmc/cb_capt.asp

Finally, before you send your final package in for processing; take it to a sea school for review. If they catch only one error, it will be worth weeks of lost time processing your license. Good luck.

To Review getting started:

First, contact the U.S. Transportation Security Administration at **www.tsa.gov**.

To obtain your Merchant Marine Credentials (MMC). Then contact the Coast Guard REC center near where you live or will work to begin your application at:

http://www.uscg.mil/nmc/cb_capt.asp

Chapter 3

Finding Your Boat

"You are not going to find the ideal boat. You are not even going to have it if you design it from scratch yourself."   Carl Lane

Captains Working Their Own Boat or Working for Others

CNN's Money Series profiled several six-figure jobs, one of which was yacht captain. Unfortunately for us there are not too many of those six-figure jobs around, particularly in today's damaged economy and if you do find one you will have spent several years working your way up or perhaps marry your way into it.

But, if you truly enjoy people, are free to travel and can take the necessary training to handle big boats, you can earn that big income on the water. Interestingly, some yacht captains earn more than ship captains of large freighters, bulk carriers or tankers.

In some large yacht jobs, you are more of an executive than a boat captain. To be sure, you make the decisions regarding the safety of the vessel and her crew, but you can spend a lot of time "managing people" rather than driving the boat.

You may spend time just moving the yacht from one location to another with minimum crew aboard, followed by a couple of hectic days cleaning everything to a white glove level before a charter party or the owner comes aboard.

Of course, you are on duty most if not all of the time. Only if the yacht is laid up or in a yard for maintenance does the captain really have time off. Most captains I know are never "off." They worry about their boat no matter where she is.

When looking for a job on a boat, get out and make connections. Go to the waterfront where you want to work and talk to everybody. Ask who might need crew; lend a hand to help other captains. Get known.

Take any delivery crew jobs offered and deliver the boat quickly and safely. Ask for a reference when you finish. Accumulate comments and references from all the skippers you sail with or owners you work for. The more experience you get on other's boats, the better prepared you will be when it's time to run your own.

Keep a personal logbook of all your time spent on the water. Even if you only spend time cleaning fish or varnishing brightwork. In particular, keep track of all trips you make, whether for pleasure or for pay.

From where to where, time en route, time of the year, sea miles (nautical miles), what vessel and who was the captain. (Get his address and phone number; you may need him to verify your sea time.)

Also, note any unusual weather you encounter, such as a hurricane passing nearby. Name the storm and give the dates you were involved with it. Someday you may be nosing around for a position on a boat and having a log showing the storms you weathered and trips you've made may weigh heavily in your favor. You will also need the sea time to document your Coast Guard license application and upgrades, your logbook will prove invaluable in documenting your time.

Talk to yacht brokers in your area, tell them what you are looking for, follow up and keep in touch with those who have an interest in you. If you find a friendly broker, stay with him, do some favors, get noticed.

Then when October comes make sure you are in Ft. Lauderdale for the greatest boat show on earth. This is where many boat jobs originate. Maybe you can get a job helping to deliver a boat to the St. Thomas Charter Boat Broker Show. Following St. Thomas is the British Virgin Islands Charterboat Show in Tortola, which in turn is followed by the Antigua Charterboat Show.

Hang out on the docks. Look and act competent, be responsible. Don't be a pest but offer to help someone get their boat ready and make a few friends in the process. A helping hand is known. Talk to every broker and sales representative who will listen.

Give them your business card (with your color picture and cell phone number on it) and focus on boats in the range you are looking for. It helps if you are pleasant to look at and have a ready smile and easy personality. A good attitude helps. Did I say it helps? It's damn near critical.

Long, unruly hair, hardware puncturing your face or body, and tattoos that attract attention will not, repeat not; get you a job on a decent boat. Smoking is often prohibited aboard the better charter boats. It's a matter of cleanliness and fire safety at sea. If you smoke, you might want to reconsider this downside of your habit.

If you use recreational or other illegal drugs, even occasionally, forget working in the charter boat business. No captain will run the risk of having his boat confiscated by authorities in some out of the way place because someone in his crew was smoking dope or worse. Many island societies are deeply religious and offensive behavior in their eyes can mean serious trouble. In truth, there is not much room for bad habits of any kind aboard any boat.

Sailing Charters

Crewed charters usually operate through a broker or clearinghouse who keeps the calendar for the boat and, in essence, tells the captain who is booked, who he has coming and where and when to pick them up. The broker keeps 10-to-15% or more of the charter fee but in return they do much of the paperwork and advertising for clients.

With the internet and today's cell phone technology, many captains operate their business without a clearinghouse or broker. However, the captain (or his mate) will spend a lot of time chasing business. Get aligned with a good, active broker. He or she will save you a lot of time and money trying to make the sale. (It is said, nothing happens in business until someone makes a sale, and that is very true in this business.)

A good captain and crew team can make a week aboard a sailing yacht one of life's truly great pleasures. The customer gets to sail on a big, powerful ocean racer or cruiser to his heart's content, safely, while his family or mate is catered to and pampered. On a good charter, everyone finds something they like to do. Swim, help raise sails and steer, snorkel, dive, fish, just read or go beachcombing and sunbathing.

A friendly, helpful captain and mate can have a good time with their guests also and enjoy being on the water for days on end with new interesting friends. I've met people chartering who became lifelong friends. Some have high visibility jobs and it's fun following their careers later in the news. I had the man who invented fiber optics aboard for a week, then best selling authors, TV entertainers, and many self-made entrepreneurs. I count my years sailing as some of my finest. What better job could there ever be?

Once a retired Navy captain came aboard who had spent most of his career on the bridge of ships, giving orders. He had never actually steered a vessel. I offered him the wheel our first day out and he never let go. He was the helmsman on our charter yacht, all day, and every day that week. He loved it, we loved it, and his wife said he never looked so happy.

Motor Yachts

Motor yachts today tend to be larger, 80' to 150' and up and usually have a permanent crew of at least three, sometimes more. The captain has more responsibility and he and his crew must provide the charter guest or owner with impeccable service, outstanding food, good company, water toys for activities and local knowledge of every port and anchorage they visit.

A good captain will also know the names of the dockmaster and dock hands in ports and marinas where they call, thereby insuring good service, prime slip locations, and a pleasant experience for his guests. Take good care of the people who serve you ashore. They are an important part of your "extended" crew.

Very few large motor yachts are owned by their captains. They simply cost too much for the average captain to buy and maintain. Wealthy individuals or the corporations they control, own most large yachts.

These owners charter their yachts for income tax breaks, as tax shelters, or simply to offset some of the cost of maintenance.

A two million plus dollar yacht grossing $300,000 a year in charter income, (ten weeks at $30,000 or some combination of short trips and full weeks) will spend $150,000 or more in wages for captain and crew and another $100,000 in maintenance costs, just to keep the yacht pristine. Then every few years the owner must find another million or so for a major refit and pay for constant yearly upgrades just to stay competitive.

The cost of wages and maintenance offset the boat's income so the owners pay a small amount in taxes on a very small profit, if any.

There is not enough money left after maintenance and labor costs to support the purchase and mortgage of a multi-million dollar yacht. Hence, most yachts have substantial financial backing before they ever board their first charter.

Some yacht brokers will be honest with prospects who want to own a charter yacht and show them that less than 5% of charter yachts ever make a real profit. However, many people of means have the dream of owning a yacht and chartering it helps them offset some of the cost of maintaining their dream.

In the large yacht industry now there is an overproduction of large yachts and not enough buyers moving up from the smaller boats to take the surplus trade-ins. Worldwide there are thousands of yachts, one hundred feet or more in length and more are being built every day. Ten years ago, yachts in the seventy to ninety foot range were considered "large" boats. Today you must have a boat over one hundred feet even to be noticed in some marinas. While this means prices are dropping for the smaller used boats, the cost of maintaining these smaller monsters is not, repeat not, falling. In fact, the older a boat gets the more it takes to maintain her in top shape.

The charter industry is very fussy about chartering older boats, even if they have recently been upgraded and rebuilt. There are just too many new boats to pick from and too many new owners of large new boats scrambling for the charter dollar.

If you are considering purchase of an older power yacht to charter, consider very carefully if you are placing yourself at a disadvantage right from the start. Talk to several yacht brokers before you make the commitment. The one exception is if you are considering a classic pre-1950's yacht that has always been properly maintained and still has her charm and beauty, with a sound hull and machinery to match. Classic yachts are enjoying resurgence today and are eagerly sought by collectors and savvy captains.

Owners Who Hire Captains

The best captains are truly "Renaissance Men" as in the age of enlightenment. Back then, a cultured gentleman was expected to have mastered several languages, be knowledgeable in scientific and medical matters, have historical perspective, and be proficient in the arts, perhaps playing some instrument agreeably.

I actually know and admire several captains who fill that bill. They are world traveled, speak a smattering of several languages, and can converse adequately with customs officials in the local language in countries they usually visit. These captains are knowledgeable about local attractions, know what is locally available in fruits, vegetables, and wines and know how to serve an elegant dinner aboard, perhaps using fish caught by guests that afternoon. Then after dinner, these captains may entertain their guests on the guitar or other instrument appropriate to the setting.

If this seems like an impossible task, consider then this captain may also have spent the morning tearing down and rebuilding a generator fuel system clogged by dirty fuel, or clearing heads plugged by unthinking guests.

Meanwhile, he is cajoling crew back aboard after a night on the town where he still has to settle a bill from an angry pub owner over broken furniture. Then he gets a forwarded mail package that informs him he has to bone up and take a course on new MCA, SCTW or IMS standards course to satisfy the yachts insurance carrier for an upgraded license.

Sad to say, the days of the rum drinking, bare-chested rollicking captain of old are long gone. Captains of large yachts draw their salary from an owner, but actually represent his insurance company's interests. When you get to the mega yacht stage, controlling insurance risks are one of the captain's prime concerns.

An inconvenient 'soft' grounding in a smaller Hatteras might result in a day's delay for an underwater prop change and a total bill of five to eight thousand dollars.

On a mega yacht (over 100') the same 'soft' grounding could cost thirty to fifty thousand for new props and an expensive haul out not to mention the lost income at the height of season at perhaps $30,000/week.

Today captaining a yacht can be brutal work, always on call, keeping the boat spotless and ready to go anywhere, anytime to meet the owner's schedule regardless of the local weather.

Keeping the crew happy and willing is also a major responsibility for the captain for he must depend on them sometimes more than they know.

Then captains must sometimes deal with owners who will not spend the necessary maintenance money to keep the yacht in good operating condition, much less pristine. See Chapter 11 for more on Owners & Captains.

An active charter yacht may do ten to twenty weeks a year on charter, add to that travel time, maintenance time, get ready and clean up time and neither you or your crew have much time for yourself.

Some Captain-Owners who have an unusual or historic vessel have developed a following from years of successful chartering and go on happily chartering year after year with repeat customers and new clients.

Few captains remain with the same owner over the long haul however. Boats are sold, the owner's interest in boating goes up or down and often when owners' get new wives, the new wives have their own priorities, which may not include a boat other women, have been aboard.

Then, captains just burn out. They want to spend more time with their families, and decide to do something else that keeps them closer to home.

On the happy side of this picture, I know one captain and his mate/wife whose owner sold his boat after many years of chartering but the owner continues to pay their regular salary as they retired to a wonderful house in Antigua. Another captain served his owner for years and when the new wife said, "Get rid of that boat!" He simply gave it to his captain. However, these are exceptional owners.

Bareboats

You are probably familiar with the term bareboat as it applies to chartering a boat for a few days or a week in some popular sailing destination like the Bahamas, Virgin Islands or

Florida Keys. Many of these boats are too big for the typical "charterer" to handle with his limited experience so charter companies sometimes require a captain to be aboard, paid for by the charter party to insure the safety of the people and the boats. Often this is only for a day or two to check out the charterer and sometimes it lasts for the length of the trip. Often too a cook is brought along on the charter as well (what wife wants to cook on vacation).

These jobs are a good place to get experience on the water and do a lot of sailing on different boats. The pay is marginal and sometimes the locals have all the business locked up. But try talking with charter companies about their need for a captain on call.

You might do some boat deliveries for them to establish your credentials, and then when they need a captain in some location, you are a known quantity.

The other call for bareboat captains is in the large yacht industry. Typically, if a company or individual wants to charter one-hundred footer or more, they will contract with the yacht owner for the yacht only (a bareboat charter.) The individual then selects a captain from a list provided by the boat owner or broker to run the boat for them. Usually a cook and other crew are contracted for as well. There are several advantages to this, since it is a private charter, i.e., they "own" the boat for the period of the charter and as such can pack aboard up to twelve family members and friends.

The Coast Guard sees this as a "bare boat" charter, not carrying passengers for hire. Off the charter party go on holiday. They go where they like, spend as much or as little as they like and have all the advantages of owning a large yacht without all the headaches.

The captain is someone acceptable to the yacht owner (generally his regular captain) so he knows the boat and the cruising area. Again, you need experience to get these jobs, but they are a place to start. Contact charter brokers or boat owners when you have your license and are ready to work.

Dive Boats

Dive charters exist almost anywhere there is clear water and something to see. Diving takes dedication, strength, stamina, and most dive masters are a hearty breed. Dive boats often schedule three dive trips a day, morning, afternoon and night dives. The wear and tear on boats, equipment and divers is significant.

Usually you will find a lot of competition around the good dive spots, which hurts also, making this one of the less lucrative boat businesses. But then again, most people in the dive business are there because they love diving. They would do it anyway.

There should always be a licensed captain (another job opportunity) aboard the dive boat when divers are in the water. Never cut corners by being the dive master and captain simultaneously, it has had fatal results too many times. Always make sure you have all divers back aboard before leaving a dive site.

A recent stranding of a diver offshore in California and the subsequent movie will affect diving negatively for years to come.

Glass Bottom Boats, Sightseeing, Water Taxi

In almost every harbor there are individuals making a living filling a local need. Glass bottom boats take out tourists and locals alike; at Wakulla Springs in Florida, many spots in the Florida Keys, several Bahaman locations, in the Virgin Islands and in Hawaii.

Capt. Bill was a popular figure on the Charlotte Amalie waterfront for many years with his tiny fake riverboat glass bottom boat. Every hour he scared his boatload of tourists by motoring over very shallow coral reefs with colorful fish hanging around waiting for his handouts, which he surreptitiously dropped off the stern as the boat, slid over the reefs.

Almost every harbor with any tourist business at all has at least one tour boat taking passengers for a boat rides. Visitors get to spend a little time on the water, perhaps take a longer tour with a boxed lunch or an evening cocktail cruise. These boats fill a need and provide steady income for their owners and crew if they have a good market and deliver a good, safe time on the water. A United States Coast Guard Master is operating every one.

Deliveries

Yacht deliveries are another way to earn your keep aboard and get some sea time. Delivery captains making long distance deliveries generally earn two to five dollars a mile plus expenses and return airfare.

You can charge more if the job is difficult or the vessel is valuable. Crew is usually paid a dollar or two less than the captain is. This is not a lot of money when you consider the hazards involved in taking an unknown, untried boat with possibly a pick-up crew sailing to an unfamiliar destination. Some captains only make deliveries for boat manufacturers and bare-boat companies. That work generally pays less, but the captains work more consistently. They also have the advantage, if you can call it that, of sailing a new boat. In theory on new boats everything works as it should, but it rarely does.

New boat or not, you will likely spend part of the delivery fixing or jury-rigging something or other. Another drawback is that you are usually on a schedule and have to go regardless of the weather. If you have to wait for days in some port for spare parts, crew, or the weather to improve, you're losing money. Lay days simply dilute your earnings.

Deliveries have other benefits though; you sometimes get to sail into new harbors and hang out in waterfronts you would never see otherwise. Keep a good log with lots of details and you can turn it into a book of your adventures one day.

John Kretschmer's first career was writing, and then he switched to delivering boats. He's now back writing and has a wonderful sea bag of stories to draw on. Read Flirting with Mermaids, for some great delivery stories.

Written between the lines, however, are days of being alone while the rest of the crew (if you have one) is asleep and you're alone on watch.

You lose track of family and friends, have few possessions and build up few resources for your later years. With no health insurance and no doctors nearby you learn self-reliance in many things. Still in all, being a thousand miles offshore on a great sail has its own rewards, which you only learn when you get out offshore. Go offshore if you can, you will never forget the experience.

Racing

Racing sailors seem to get their jobs by coming up through the ranks or being related to the owner. Kids often get started in racing with one-design boats and move up to larger boats as their age and skills progress. Eventually they are noticed and be asked by other racing skippers to help crew their boats.

In the beginning you are lucky to get expenses, later as your experience grows and you are in demand, you may work for a salary by whoever is sponsoring the boat. Many racers simply do it as a hobby with no thought of making a living. Without a silver spoon firmly clenched in your teeth, racing as a profession seems tough at best.

In Michigan, Joe Dissette built his hobby of racing Lightning Class sailboats into a lifelong business of buying, rebuilding and selling Lightning's, winning several National championships along the way. Joe once sold a lightning to a customer in Ireland and took his wife over there to lend his expertise to an Irish Lightning Club. He says his team won all their races but it just may be "blarney."

Fishing

If fishing is your passion, a sport fishing boat offering a half or full day fishing offshore might be your ticket. Most boats take six or less anglers and include a mate to bait hooks and generally keep the customers supplied with beer and jokes.

Usually these boats are found grouped together along a stretch of waterfront, very visible and accessible to customers. If there is a lot of competition, you will need some means to make your charter operation stand out. A new, clean boat with good gear will be noticed. Then all you have to do is find the customers and the fish.

Like anything else, do what you know. If you know an area like the palm of your hand, know the water, the reefs and shoals, sand bars and deep holes and read everything you can about fishing; then you will probably be a successful fishing boat captain.

If you are new and green to the area, it's better to first take a job aboard another boat until you've learned all you can about the area both under and above water. Be nice to your captain-teacher, stay his friend after you go into business.

Big money sportfishing tournaments are the ultimate sport if you are into that sort of thing. Some of these boats chase the competition from tournament to tournament, sometimes even shipping the boat overland or overseas to fish. Boats fishing these venues must have the best gear and most knowledgeable skippers and crew. The stakes are very high as are the expenses. Seldom do the skippers own these boats.

It takes too much capital to follow the tournaments unless you have another source of income to finance your passion.

You can probably afford to use your own boat for inshore or back bay fishing if it's a small boat or "flats" boat. Fishing guides whose customers catch fish generally are successful if they are out a lot and control their expenses. This seems obvious; however, it is a hard life, in the sun all day, chasing the elusive catch, fish and clients.

On small boats where you are the owner/captain, when you don't work, no money comes in and you can hit an income plateau here very fast. You can only work 365 days a year. Whatever rate you charge for a day times the number of days you fish, that is your maximum income. Busy captains seem to do all right, but they must work every day they possibly can finding customers and fishing.

Another type charter boat/fishing guide operation popular is freshwater fishing. These captains pick up their customers and then trail their boats to a particular area, lake or river where they launch and speed off to the area picked for the days fishing. Once there, they may troll with an electric motor or pole the boat through shallow water, seeking the elusive fish-of-the-day.

These fishing operations depend a lot on word of mouth advertising, rack cards and Internet pages. With just a cell phone and some advertising, the captain-fishing guide is in business. Licensing is usually simpler also, just a six-pack license, a business license and a fishing license.

Remember though, even white water rafting requires a license if you charge for your services.

Commercial head boats that take 20 or more fishing offshore for half or full day of fishing require Coast Guard inspected vessels and a big investment in gear to get started. You may work up to this business or work for someone else, every boat needs a captain.

You might dream up other combinations to entice customers, perhaps a week's cruise onboard your pleasure yacht stopping along the way at various fishing holes and desirable anchorages you've scouted out. There are endless possibilities, and only you will know what works for you and your customers.

Motor Boat Rides

What probably got me started in this business was a ride in a classic Chris-Craft runabout when I was about three or four years old. An old man gave rides in his boat on the Elk River in Noel, Missouri for a dollar a head. My Dad would buy me a ride every time we passed through Noel. I can see that boat to this day-in my head. You might start some kid in that direction with your boat. Teach him well.

Buying a Sportfishing Boat

If you are considering buying a sportfishing boat and then chartering it for a living, proceed very carefully. First check out what the competition is using. Chances are, if there is a charter business locally doing what you want to do, they have figured out the suitable boat, perhaps the hard way by trial and expensive errors.

Look carefully before you try something very different. Your new operation could have an advantage with a newer model, more speed or better sea keeping abilities. Of course this will cost extra and you can only justify the added expense if it gives you enough advantage to charge more to cover the extra cost.

Generally, fishing charters concentrate on the fish, the boat is relatively unimportant as long as it can get to the fishing grounds quickly, looks and is seaworthy, has a comfortable ride, and is rigged adequately for the fishing.

Of course having added amenities like air conditioning and heating, an adequate galley, decent heads, and other creature comforts for the fishing party and perhaps wives or others not actively fishing will make the day more pleasant for all aboard. Most importantly, the boat must be clean, be safe and be seaworthy.

Gas powered boat engines tend to use a lot more fuel and wear out sooner than reliable diesel power. Make sure your boat has sufficient fuel capacity to stay out all day with an adequate fuel reserve to get home safely. Expecting to compete with other professionals with an old tired pleasure boat, one without the latest fishing equipment or a boat with a too small fuel tank that restricts your time on the fishing grounds is foolhardy. Your business will struggle from the beginning.

A boat that is perfectly fine for you to go out occasionally and troll for the "big ones" may not have the speed, endurance or reliability to stand up to fishing day after day in every kind of weather. Many pleasure fishing boats simply will not work.

Unfortunately, many boat owners find out too late that big gas engines in an otherwise fine-fishing hull, drink all their potential profits at the gas dock. Check any boating magazine; there are always bargains in fine hulls with gas engines, sometimes half the price of the same boat with new diesels. Buy a gas-guzzler and you will pay and pay.

The maintenance cost of gas engines seems to run two or three times that of a simple diesel.

You need an engine that is fuel efficient, reliable and easy to service. Stick with something simple that runs day in and out requiring only fuel, oil and filter changes.

Buying a Luxury Yacht

Yachting charters aboard large power boats is a different story altogether. Here luxury, the more the better, is the package. You will provide great food, comfortable surroundings and a pleasant attentive crew for starters.

There must be something for everyone to do. New water toys, fishing gear, snorkeling, dive gear, hot tubs, and entertainment centers are necessary along with high tech communication gear for worldwide communication and internet access.

A well stocked DVD library with big screen TV and perhaps one in each stateroom as well. Have I left anything out? Whatever some charter guest wants, there is a boat somewhere that offers it.

After a few years, sometimes a few short years, owners trade up and large luxury power yachts come on the market at bargain prices, or so they seem. Large power yachts require deep pockets. Their annual maintenance cost is high, think hauling an eighty footer, buying bottom paint and then paying someone to put it on, no more do it yourself here.

Consequently, these boats eat a lot of cash up front and the business is subject to many variables. The cash flow can be impressive, but the costs are also.

Classic Yachts

Chartering on classic yachts, power or sail, is a popular alternative to the usual new, fancy yacht charter. Several classic yachts that have been restored to better than new condition are now plying the world's oceans, up and down the ICW, the Great Lakes and western waters. They are providing their owners and guests the very best of yachting, much as it "used to be."

The Classic Yachting Association (CYA.org) membership is growing with classic boats and those who appreciate the classic yachting period, that time between the two great wars, charter many of these boats for the experience without the expense of ownership.

Classic sailing yachts like, *Gleam, Columbia, Easterner, Heritage, Weatherly, Onawa, American Eagle and Intrepid* take groups of twelve out for day sails and match races in Newport. The sight of a half dozen ex-America's Cup boats tacking upwind to a marker is reason enough to be in this business.

I owned a classic wood yacht for several years and restoring and working her in the charter business was very rewarding. Every day was another labor of love. Hanging out on the boat talking to people on the dock who are attracted to classic yachts is great fun.

Locked up inside many otherwise ordinary looking people are sea stories of their adventures with the old classics, sail and power, large and small. It was good working her and good for the boat. Boats need to work is my philosophy. (And people too.)

Working Your Boat

The reason you choose this path is that you want to enjoy working on the water and make enough money to keep doing it. Like other captains, you've probably worked at other jobs for a long time. Family, friends and obligations all pile up and then one day you've had enough. You just want to break away, to do what 'you' want to do.

Warning! It is very easy to hurt the ones you love following 'your' dream. A dream can be a selfish thing and, let's face it; men usually get the itch of chartering for a living about the time they figure out their job doesn't really satisfy their need to control their own life. Sometimes their marriage begins to go bad and, living and working on a boat, without all the accoutrements of land living, seems a refreshing thing.

It happens to women also, I know several who left a 'perfect marriage' to go sailing. It was their passion and the hurt that follows can cut both ways.

A lucky man or woman finds a partner early in life with the same yearnings for independence. They share a desire for freedom that drives them to work as a team for their charter business. Most partnerships that share a common goal are very successful.

Unless you have enough money from some other source to last the rest of your life, you must be successful at chartering to be able to continue doing it and having fun. It's no good struggling financially month-to-month, hoping the business will take off and you can get back to doing what you like, namely running the boat.

Use the information and tips in this book as extra leverage for your business, the edge to help you follow your dream.

Few charter skippers are independently wealthy. Although some skippers have made serious money, sometimes even making that money chartering. Most don't. They fail to see that chartering is really a business, a business that must take in more than it spends.

I hope this book will help you see yourself as a businessperson as well as a captain. Be the one who makes his or her living on the water, and does it well. But then if you only want your chartering to supplement your income, perhaps help with the costs of owning your dream boat, that's all right too. There is nothing wrong in making a few bucks on the side to help with your expenses.

Look around any fancy marina and you will find the men and women operating the big million dollar boats are, for the most part, hired captains and crew. They have a limited life of their own, always at the beck and call of a corporation or private owner.

Professional captains often move from place to place serving others, not unlike the servant class in the English movies you see on PBS. Captains live the grand lifestyle, are around other captains of industry and the wealthy leisure class, but are still the servants, the facilitators. You only drive their boat.

Many captains get charter clients who are rich and famous and are as easy to get along with as anyone else is. Some have inflated egos pumped up by their handlers and fans, but even those super stars have to be told where the head is and how to work it.

So mostly it comes down to treating everyone the same, i.e., with respect and courtesy. And if some spoiled brats, no matter their age, trash the boat, well, usually their handlers will pay and you get to redecorate the boat.

Some other charter captains are just able to squeak out a living using their own boat. Chartering occasionally provides enough money to maintain a boat and a lifestyle that others envy. If this is you and you want to earn more and live better, use this book to make a difference in your operation.

So, a final word of advice; start small, get your feet "wet" and work your way up. You will enjoy the trip and your chances of success are much greater.

Documentation, Registration

Your charter boat will be documented, registered in your state, or registered in a foreign country. This documentation or registration proves who owns the boat and what your boat is licensed to do i.e., (pleasure use, commercial use, government use, and where it is authorized to operate).

You have some choices in whether to document with the Coast Guard or to register with your state. The choices will depend upon the size of your boat, where she was built, how you intend to charter or rent her out, whether you are a U.S. citizen and how much money you want to spend on registration or documentation.

State numbering (registration) is the simplest and usually the least expensive. State registration will generally restrict you to carrying six passengers or less. Most states have a title law as well as registration laws and if you live in one of those, you can title your vessel and obtain registration numbers at the same time. State registration is usually sufficient if you are not planning to carry more than six paying passengers at a time or travel to a foreign port, i.e., the Bahamas, Mexico, Caribbean, or Canada.

To carry <u>more than six for hire</u> or to go foreign, your vessel will need to be documented and inspected. A Coast Guard inspection officer or team will inspect your vessel and assign your vessel a number of passengers your vessel is registered to carry.

If you document your vessel, the documentation certificate issued by the Coast Guard will be endorsed for fishery, coastwise trade, registry, (foreign trade) or recreation. Your private pleasure boat will probably be registered for recreation. Any documented vessel may be used for pleasure or recreation regardless of her endorsement. You can take your family and friends out for a picnic on your tuna seiner if you want.

A person (natural or corporate) may charter a vessel documented for recreation use, as long as there are no more than twelve passengers aboard. It must however be a bareboat charter, that is, the charter party supplies the captain, the crew and pays all expenses of operating the vessel (food and fuel). Certain large yachts may carry a certificate for up to 30 passengers however; they are built to much stiffer standards than your yacht.

Coastwise or Registry Documentation

If your documentation is coastwise or registry; you can sell tickets to the public and run the boat yourself (with a license) or with a qualified captain. The documentation and examination for coastwise or registry by the CG will determine how many passengers your vessel can take. (Remember the discussion under Chapter 2, 100-ton license about inspected vessels.)

You cannot sell individual tickets on your recreational documented pleasure boat to twelve passengers and take them on a cruise or tour. You can however, sell tickets to six passengers and take them out, perfectly legal.

However, with a recreational endorsement, owners of larger yachts usually charter the whole yacht to a single charterer. (Who brings along his family or friends up to twelve and supplies the captain and crew.) The owner supplies a list of captains known and acceptable to him for the charterer to engage. That way even though it is a bareboat charter, the owner still has some control over his yacht.

Big boat charter brokers have this down to a fine science and you can usually rely on their expertise but, of course, it will cost you in broker fees. If you have a large documented yacht and wish to charter her for more than six passengers, talk to a known charter broker in your area. You need local expertise in this area.

The very first U.S. Congress authorized documentation for ships in 1789. It primarily provided conclusive proof of nationality of a vessel engaged in commerce. In 1848, documentation was expanded to include pleasure vessels to avoid them having to check in with U.S. customs when entering and leaving port on pleasure cruises. At that time the yachting ensign was authorized to distinguish pleasure boats from commercial boats.

The Yacht Ensign

Is the familiar 13 stars surrounding a fouled anchor in the blue field of our national flag is called the "Yacht Ensign," It is now allowed to be flown from all recreational vessels, documented or not within the U.S. and its territories.

If your vessel documentation is <u>Commercial, Fishery, Registry, Coastwise or Government</u> you may only fly the U.S. Flag (Ensign).

If you take your U.S. vessel into foreign waters, all vessels must fly the U.S. Flag since the yacht ensign is not an official flag of the U.S.

Since 1920, documentation has also allowed for preferred mortgages on documented vessels, making the financing of yachts practical (and contributing to the worldwide glut of large, overpriced yachts.)

A Documented Vessel

To be documented a vessel must be at least five net tons and with few exceptions must be wholly owned by U.S. citizens. (Most vessels over 25′ in length will meet the minimum 5 ton restriction.)

Commercial vessels, no matter the size, that operate wholly within a state on state waters or within canals or harbors of a state are exempt from documentation, if they so desire, i.e., harbor tours, water taxi's and fishing trips on inland lakes, to name a few.

You may establish vessel ownership on a new vessel by either; builder's certification on a custom vessel or manufacturer's certificate of origin naming the applicant as the person for whom the vessel was built or first transferred or a copy of the state registration or title or, foreign registration, which shows the applicant, owns the vessel.

Your citizenship is established by birth certificate, passport or in the case of a corporation owning the vessel, the chief executive officer and chairman of the board must provide proof they are U.S. citizens. Corporations owning fishing vessels must provide other specific requirements.

Commercial documented vessels (fishery, coastwise trade, and registry) do not display their official numbers on the hull but instead are identified by the vessels' name on each side of the bow and on the stern. The hailing port may be any named place in a state or territory and is added to the stern below the name. The state name may be abbreviated.

Recreation documented boats are exempt from displaying the name on the bow, but the stern displays the name and hailing port. The name may be displayed on decorative name boards anywhere on the boat, commercial or pleasure.

There are specific requirements for display of the name on the bow and stern and these will be given you when you document your vessel at the National Vessel Documentation Center (NVDC) you can get more information from their website at:

www.uscg.mil/hq/g-m/vdoc/nvdc.htm

Documentation Names

Names of documented vessels may not exceed 33 letters and may not resemble any word used to request assistance at sea, be phonetically identical to obscene, indecent or profane language or to racial or ethnic epithets. Once you establish the name and hailing port, it may not be changed without application and fees to the NVDC.

There is no rule against duplication of names, so the hailing port is helpful in establishing your unique name.

The official documentation number must be carved, molded, welded or otherwise made a permanent part of the structure of the hull so that removal or alteration would be obvious and cause damage to the surrounding area. These numbers are normally located in the forward part of the vessel or in the engine room, whichever seems more logical and obvious.

The actual numbers are preceded by the abbreviation "No." and the numbers must be at least three inches high (for example No. 500555) Many vessels have more than one owner so the Coast Guard requires that one owner be designated the "managing owner" to make sure the right person gets mail or notices from the Coast Guard.

Documentation also has a side effect of allowing the government to requisition your vessel in wartime if it is needed.

During WWII, many pleasure and fishery documented vessels proudly served our country, patrolling coastlines, performing picket duty, carrying cargo between coastal ports, and some California fishing boats even carried refrigerated cargo to our forces in the south Pacific. (Imagine steak and ice cream on Okinawa for our troops.)

Other boats were used to train seamen and in one documented case, a former private yacht managed the probable kill of an enemy submarine off Miami by ramming it.

While most sub-chasers were government built boats, some were private yachts converted to military duty. Most boats requisitioned were given the designation of YP class vessels as in Yard Patrol. i.e., the requisitioned boats were assigned to a particular Navy yard and used for patrolling, coastwise or offshore.

SC class vessels were government built Sub Chaser class boats, built in both wood and steel. Sailing craft were particularly useful patrolling for submarines, as they were hard to detect by the subs and more than once a submarine surfaced practically underneath a sailboat on patrol, sometimes with disastrous results. Several well-found seaworthy sailboats were lost while on patrol duty and the belief is they were discovered by armed enemy subs and sunk.

The Patrol Craft Sailors Association, PCSA, is dedicated to keeping alive the memory of the exploits of those brave and selfless men and women who served aboard these vessels.

For those interested,"Splinter Flee" a book by Theodore R. Treadwell, details exploits of the patrol and sub-chaser boats in WWII.

During the Second World War the motor yacht *JODARO*, was requisitioned by the Navy, re-numbered YP-611 and for four years trained navy divers and patrolled the waters off San Diego. Many years later I owned that yacht and found traces of Navy gray paint in her engine room and bilges, confirming the old Navy adage.

If it moves by itself, salute it.
If you can move it, take it.
And if you can't move it, paint it.

Documentation does not relieve you from complying with your state's laws regarding registration of your vessel. Most states require that you display state decals showing compliance with state requirements. Dinghies and yacht tenders cannot be documented and if powered, fall under the owner's state registration laws. At this time, the documentation must be renewed annually by mail, but there is no fee for renewal.

Owners of foreign built vessels (Formosa, Canada, Norway, etc.) may now be able to use their vessels to carry passengers by obtaining a waiver from:

Maritime Administration, Small Vessel Waiver Program, MAR-830, Room 7201, 400 Seventh St. SW, Washington, DC 20590.

http://marad.dot.gov/

The waiver allows eligible vessels to carry up to 12 passengers for hire under bareboat charter rules providing they are at least 5 tons net, are more than 3 years old and foreign built.

Finally, if your vessel or your proposed business plan does not fit any of the above and you are denied documentation, contact your congressional representative. Many boats have had special exemption riders attached to bills to enable documentation for special cases. It is not easy, cheap nor quick, but it may be your course of last resort.

Chapter 4

Your Charter Plan

"Voyager upon life's sea; to yourself be true, and what'er your lot may be, paddle your own canoe."   Edward Philpots

Making Dreams Come True

Your Charter Plan is simply a business plan to help you make a success of your business and insure it makes enough money to survive and support you. The purpose of a business plan is to focus your attention on the business in such a way that you have a clear path to follow.

Then when the financial seas get rough and you struggle to pay your bills, review your plan to see what corrections need to be made to get you back on course. You have a course line against which you can measure your drift.

Get a three-ring binder. On the first page, write your goal in starting your charter business.

Answer the question, why are you going into chartering? Is it to make a living on the water, to go fishing every day, or what?

Your "why" answer can be things like, going cruising, buy a bigger boat, retire on the water, whatever works for you. Keep your notes about your goals short and specific. Write from the heart. Be honest with yourself, what do you really want out of this?

Keep this binder handy, put in it anything you want to remember or check on. Let it be the rough deck log of your new business. (no binder, no problem-use the last pages of this book for your notes)

Next, write a fifty-word description of what you are proposing to do. Here is my fifty-word description of what we offered for river tours.

Take a luxury tour aboard a classic 60' antique yacht, up the Caloosahatchee River past towering oaks, Spanish moss, ox-bows in the river, birds, and exotic animals close at hand as we cruise regally past farmers homestead houses and new million dollar homes. Lunch prepared on board by our crew.

Notice the sentence structure is choppy because I've left out 'soft' verbs in the interest of brevity. Now look at the fifty words again with <u>underlined</u> key suggestive words.

<u>Take</u> a <u>luxury tour</u> aboard a <u>classic</u> 60' <u>antique yacht,</u> up the <u>Caloosahatchee River</u> past <u>towering</u> oaks, <u>Spanish moss</u>, ox-bows in the river, birds <u>and exotic animals</u> close at hand as we <u>cruise regally</u> past farmers <u>homestead houses</u> and <u>new million dollar homes.</u> <u>Lunch prepared on board</u> by <u>our crew.</u>

Nearly every word is pulling you into the "experience." Words like, classic, antique, yacht, towering, moss, river, birds, animals, cruise, homestead, million dollar, lunch and our crew, all conger up the idea of a "real experience in the Florida of old," or Old Florida River Tours which just happens to be the name of our business also.

Make your fifty words work very hard. When you are happy with your words, write them again leaving out every weak word that does not advance the concept, the story of your business.

After you are happy with those fifty words, <u>condense your fifty words to no more than twenty words</u>, twenty of the most powerful words you can think of that describe your business.

Keep both your fifty and twenty word descriptions; you will use them many times later in your advertising. Write them in your notebook in pencil so you can come back later and 'tighten' it up.

I cannot stress enough the need to really focus on a description of your business and to capture that focus in as few words as possible. You need that focus to project yourself to your customers and to remind you what you are doing and why you are here.

Now while this is still fresh in your mind, write a one-sentence description of what you do. So when people ask, you are ready with a quick, definite answer. Something like:

- We catch striped bass off Montauk.
- We go whale watching off Nantucket.
- We fish where Hemingway fished.
- We offer Bed and Breakfast on a Yacht.
- We cruise the lower keys, chasing the sun.
- We cruise an antique yacht to Old Florida.

Next, try to determine who your customers are going to be. In your notebook on another sheet of paper list the kinds of people you are seeking and everywhere you might find them. Will they be tourists, senior citizens, fishing buddies, locals. Who? Can you reach them with just a sign on your boat or do you need to go through a yacht broker, print rack card advertising, focus on the Internet or by personal contact?

Try your relatives and friends; do they belong to some organization or club that would be interested in taking your trip? Talk to everyone you know, ask them if they will help you get your business started. Use social networking sites on the internet, anything that works.

On that same piece of paper, write how you expect to reach your customers. Will you use newspapers, flyers, rack cards, magazines, web pages, search engines, posters, handouts, friends or word of mouth? Write your plans here. If you know the costs for this type of advertising, write them here also.

On another sheet, in front of a pocket to hold brochures, write down everything you know about your competition. Get their brochure or rack card, study it and then describe their operation in a few words. Things like; their type of boat, how many people they take, where do they go?

(It might be a good idea to buy a ticket for their trip and take it.) You will learn what they do well and perhaps what not to do.

Get prices your competition charges and write them down. What will you charge? This may be something that will be trial and error for the first year or so. With pricing, it is better to start out a little low, equal to or slightly less than your competition. Try to be in the middle on pricing until you get the business established. You want to be low enough to stand out and attract new customers but not so low, you are losing money every time you leave the dock. Leave room to adjust your prices, up or down.

If you have no competition, or none nearby, you will have to establish your market. In some way this is more difficult than being in a crowded area with lots of other competition. For one thing, you must educate everyone about what you do and why they should do it with you. This is where your communication skills become important. Chapter 6 will help you with this.

It is not a good idea to put prices on your printed material until you are established and have tested the water thoroughly. You can always have stickers printed with prices and put them on your printed material a few at a time. You can even print stickers yourself on your computer, get Avery® 8255 sheet labels and print stickers, as you need them. Then if you need to change prices, you won't be stuck with thousands of rack cards with the wrong prices on them.

Then on another sheet in your notebook, jot down how much you have in your budget to spend for marketing your boat and services.

Break it down by quarters, (three months at a time) so much ahead of your season, so much during your season. Perhaps your area has no season; it's just busy all the time.

If you are new to marketing talk to your printer and your Chamber of Commerce. They will have some good information you can use about advertising options. There are books on guerilla marketing to help you make a splash with little cash. Try Amazon or your local library for marketing books.

Now come the interesting parts. List what you must do to get you to your first day of charter. This becomes your "Get List." Write down your gets, number them and get started. Under each major heading, list all the gets you need to make each one happen. This list will grow, but keep it manageable with sub headings and related gets grouped together.

Suggested List of "Gets" to start.

- Get Coast Guard License
- Get business license

- Get boat
- Get Insurance
- Get Crew
- Get operating cash
- Get advertising in place
- Get supplies
- Get-whatever else you need to reach your first day of charter.

Next, make a timetable for the "Gets" above. Date your gets, when do you need to have them complete. Get everything you can think of and put it on a calendar. Make a timeline, if that will help, but make a "Get" schedule on a calendar and stick to it. Your time line could be as little as a week or as long as six months or more, depending on what you need to do to be legally and physically ready.

Set a start up date and try to meet it. Most businesses have a hard time opening on time. Sometimes they just have to be "forced" open. Set a realistic date and try very, very hard to meet it.

Here is a really important part, find a mentor. That is someone to mentor and monitor your progress. Your mentor could be your partner, family member or, best yet, someone else in the business. Write down the names of several possible mentors, pick the best three and talk to each of them. You should find at least one who can help you. Please don't skip this most important step. I know it is sometimes embarrassing to ask for help, but this is a very important step. This person will be your "second guesser," the person who will hopefully see things you've missed and help guide you to achieve your dream. There you have it.

How to Get Your Business Started

1. Write your goals. (make money, buy a boat, retire,?)
2. Write a 50-word description and then shorten it to 20 words.   .
3. Write a one-sentence description of what you do.
4. Figure out who are your customers?
5. How do you reach your customers.
6. Who is your competition, what do they do?
7. How much can you spend for marketing?
8. What steps to your first day of charter?
9. Make a timetable from today to your first charter.
10. Find a mentor, talk to him/her.

As soon as you can answer these ten questions to your satisfaction, you have your charter plan, your sailing orders and in your three ring binder or the back of this book, a personalized guide for achieving your goal.

It will take some time to fill in all the blanks. Anytime you get an inspiration, write it down. Inspiration and perspiration are the two main ingredients of a successful business venture.

You have a lot of research to do and please, as you go along, talk to others about your plan. Talk to your mentor, share you dreams but dig up the hard facts too. The more you know, the more you will want to know and be prepared when the right time comes.

Creating and sticking to your chartering plan is hard. Writing it all down is a pain, but necessary. (Write in pencil so you can "update" your plan as it unfolds) If you know, or can find someone in the business to be your trusted associate, mentor or coach, talk to them now.

Approach someone else in a similar business but far enough away that your business will not interfere with theirs.

Be upfront, ask for help and do what you can to repay their help. If you are serious about succeeding in this business, ask your coach-mentor to hold you accountable to your plan and your goal. Your plan will change constantly, but your goal is constant, permanent and solid. Keep it so, keep focused. If you cannot find someone in a similar business to be a mentor, don't give up. Speak to someone who is operating a successful one man (or woman) business. They will have a lot of advice for you, no matter their business venture.

If you haven't done so already, go out and knock on some yacht or boat brokers doors. Tell them what you are planning, get their feedback and offer to keep them posted on your progress of getting your business started on the water.

Many charter brokers have arrived in their business up through the charter boat ranks just like you. They will help you if they can. Your success means more business for them. You might find a real friend or mentor in the broker business. Someone who will enjoy seeing you succeed and will feel a part of your success when you are out there, captain of your own charter operation.

When you find that person, nurture the relationship, it will do you both good.

Chapter 5

The Nitty Gritty

Here are the Essential, Basic, and Specific Practical Details

*"In Nature there are neither rewards nor punishments.
There are consequences."* James Baldwin

Choosing Your Business Name is the Most Important Information in this Book

Your choice of name is much more important than you might think. Your first inclination probably will be to name your charter operation after your boat's name. Stop; think about that for a minute.

Your boat's name means a lot to you, either because you named it yourself or because you liked the name on the transom when you bought it. The name probably influenced your decision to buy it in the first place.

However, that doesn't necessarily mean your customers will have the same association and it may indeed turn them off because it's either hard to pronounce, too "cutesy" or reminds them of some negative experience.

Another reason for not selecting your boat's name for the name of your business is you may sell your boat or upgrade to a larger one and may want to, or have to, discard the name. With a unique name for your business, it doesn't matter what your boat name is. In fact you can name your boat a cute name that people remember fondly while they would probably never remember your business name fondly.

But the most convincing reason for not using your boat name is your Internet advertising. Use your business name in your Internet advertising in such a way that search engines looking for "subject matter" as in "fishing, cruising, touring, yachting, etc." will find your listing quickly. They might never pick up on your boat's name.

Search Engines are Critical to Your Success.

Your listing on the Internet will add 25% to 50% or more to your business. Read more about this in Chapter 6.

Let's look at two examples of boat names that work, then we'll look at why business names are better. In the Virgin Islands (before the Internet), we all used boat brokers to find us customers. They did the advertising in boating magazines and fielded the calls from customers.

Since many boats were owned by absentee owners and the boat captain's changed on a fairly frequent basis, the boat's name was how brokers knew us.

When I first arrived in St. Thomas trying to be noticed by the brokers, one of my first decisions was to choose ANGELIQUE for my boat's name. My reasoning was three fold.

1. The name started with an A; therefore it was alphabetized first in the list of charter boats used by brokers. (There were about fifty on the lists at that time.)

2. It was three syllables, easy to remember easy to speak and easy to understand, particularly on the VHF. (an-ge-leak)

3. Finally I believed it was a name women liked. It sounded like a romance novel. (I later learned that in fact there is a series of romance novels with Angelique in the name of each one.)

Men want a robust sailing or fishing experience aboard boats with names like Dauntless or Fish'n Fool but if the wife is going to enjoy the trip, she needs to feel comfortable being aboard. For many women spending a week aboard a boat named, WET DREAMS just won't float.

The proof of our name game was that we were busy from the start. I had other boats asking what our secret was in getting bookings, when they had been chartering longer and had bigger and better boats.

I feigned ignorance, told them it was just our good luck, beginner's luck I said. But I knew the right name was critical to the operation. It propelled us to many charters.

My next charter operation was planned from the beginning to capitalize on the name. I started a river cruise business using a sixty-five year old antique wood trawler. The boat's original name was *JODARO*. It meant everything to the original owner being made from the names of his children, John, Dale & Robert.

A subsequent owner changed it to *BLACK TIE*, an elegant, descriptive name for the grand lady she was. *BLACK TIE* fit our business concept but the name was wrong for a business name. Who wants to feel they have to 'dress up' aboard. So our business name, Old Florida River Tours contains the key elements that pull in the business and *BLACK TIE* reinforces the antique elegance.

Old Florida River Tours says what we do. A river tour to old Florida on an elegant classic yacht with an elegant name, *BLACK TIE*. Remember writing the one sentence description of your business? My sentence was just eight words.

"Cruise a classic antique yacht to Old Florida."

OK, how has it worked? From day one we had people wanting to take the cruise (tour) to experience "Old Florida." It didn't hurt that at about the same time there was a statewide backlash against all the new construction, high-rises, highways, high prices, high crime rates and paved parking lots with a sea of cars everywhere.

People want to see Old Florida before it is all gone. And we were there to show it to them. Then other companies began to jump on the wagon, we now have Old Florida Banks, Old Florida Museums, Tales of Old Florida, Old Florida Heritage Highway, Old Florida Pottery, Old Florida Festival, etc.

Do these other Old Florida names hurt us? Not in the least. Google lists over twenty one million sites with "Old Florida" in the name somewhere. Change the search to "Old Florida River" and Google listing drop to four million, add Tours as in "Old Florida River Tours" and it drops to just over a million sites with those four words in the listings somewhere with our listing being in the top five, often number one. Not a paid listing, no fancy web pages, just a unique name that tells a story. Choose your name carefully.

The Second Most Important Information in This Book

Consider the following. If you own your own charter boat, registered in your name, LEASE it to your charter business. That way if, for any reason, your charter business is in financial trouble or someone obtains a judgment against your business, you have somewhat protected your boat from seizure.

You may be able to LEASE it to another company (yours or someone else's) and keep working. Not tied up at the dock because of a court order or to satisfy a judgment. Likewise, if you own the dock you sail from or even if you rent or lease it, SUB-LEASE it to your charter business, for the same reason. You may be able to keep operating in the face of a judgment or legal action that would otherwise shut you down.

In our litigious society one must use whatever necessary safeguards to keep some frivolous lawsuit from wrecking our hard-earned business. That said, I've never had a problem by following my own advice in this book.

Disclaimer

I am only the author of this book and not a lawyer, so don't take this as legal advice. It is not. It is simply a way to help avoid some nasty possibilities in this business. Consult with your own attorney for legal advice.

Legal Name Options

If you are ready to start your own operation, decide what form of business you are going to use. Sole Proprietorship is the most common. You own the business with no other partners and you are not incorporated. If you operate your business under your own name, (Jake's Charters) the business licensing laws in most states are very liberal and simple.

If, however, you choose to call yourself by your boat name, "Rainbow Charters, Reel-em-in Fishing," or some other business name, you must register the name, with the county, the state, or both.

If you incorporate your business under "Rainbow Charters, Inc. or LLC," the corporation is the fictitious name and no further registration of your "name" is required. Go to your state website or tax office for all the information on setting up and registering your corporation. Corporation forms like Subchapter S or LLC (Limited Liability Company) pass profits through to the owner who is responsible for any taxes, just like a sole proprietor. Talk to your tax adviser on this.

Location

Location, location, location is most important, to lift a phrase from real estate professionals. There are some exceptions; term charter boats that move from location to location throughout the season, fishing boats that follow the fish or tournaments and racers. A delivery skipper located in Iowa will not get many calls from Miami obviously.

Several charter yachts operate "offshore," staying in the Bahamas or lower Caribbean and flying guests in for a week or more on the yacht for fishing or just cruising. In those instances quick easy communication is paramount, it might be necessary to obtain a satellite phone just to keep in touch and arrange pick-up's and supplies. Anything is possible, just be aware of the extra costs involved in getting clients to and from your boat and the extra cost of necessary food and other supplies.

If you are operating in foreign waters, ask around other boats and marinas for the local governments' attitude towards charter boats. Some are still just happy for whatever business you can throw the local business community, others take a hard line and want you to be licensed and perhaps pay some form of tax. Be careful. You may appear to be just another rich yachtsman passing through, but showing up in the same place too many times will raise plenty of questions you probably don't want to answer.

A variation on this is to operate up and down our coast, as several successful charter operations (and a couple of coastal cruise ships) have done for years.

They base themselves in a "home port" and schedule cruises North or South along the intercoastal waterway, picking up and discharging guests in various ports as they go along.

Finding a Good Location

My dock is in the city marina, two blocks from the heart of town. I got that prime corner slip facing the street after waiting about 5 years. You may be more fortunate, or it may be impossible to snag a good slip until someone leaves or dies. Whatever it takes, make the effort. Being where people can find you is most important.

The other large part of my business activity (other than the Internet) is from people strolling by the boat or seeing our dockside signs or who have gotten a recommendation from someone who took our trip or stayed aboard.

This encourages me to spend a lot of time on the boat just being visible while she is docked in her slip. Many a day I would rather be doing something else than working outside on the vessel. However, every day I'm out there, someone stops to talk and they walk away with an impression and a brochure....priceless free advertising.

I also have our name and phone number visible in large letters on the weather-cloths on the vessel. This has produced several charters from people calling to book a trip as we were passing by on the river. They saw us having so much fun they wanted to book a trip for the future.

If you are starting from scratch, it might be worthwhile to look into locating in an area where there is little or no competition. It takes time to build up a business location. However, if you can build your business without having to compete with others doing the same thing, your location could be just the advantage you need to succeed.

Several things you need on your dock are obviously power and water, a nice clean trash container that's kept clean and emptied regularly, a place for a very visible sign, some shade and a place to sit and chat if possible. If you can provide free parking that is a big plus. The more room you have, the more visible you will be and can accommodate large dock crowds when necessary.

Your dock should be safe to walk on, no place here for old rotten planks and trip hazards. People should be able to get on and off your boat with a minimum of effort, so no shaky steps or ladders. If necessary, build a platform or ramps to get people on and off expeditiously and safely. Buy ready-made steps for mobile homes, they are sturdy and much less costly than anything you can buy in a marine store.

Provide handrails where necessary and always be close by to give guests a hand embarking and disembarking. Take their bags, handbags, umbrellas, cameras, and anything else they are carrying so their hands are free to hold on. An accident boarding will ruin their day, your day, your reputation and possibly your bank account.

If the way out to deep water is difficult, shallow or torturous, practice enough times by yourself so that your can make the way out and back in seem effortless. Confidence in the captain goes a long way in making the trip a wonderful experience for the guests. They do not want to know how you sweat to get them out and in safely.

Give polite instructions where you want guests to be when you leave your dock and when you arrive back at the dock. Some people will be in a great rush to get off or will try to help and just get in the way and might get hurt.

Lastly, stay current with your landlord or Dockmaster. You will need their help and encouragement to succeed. Pay your dockage on time and keep your area clean and neat. Be Mr. Nice Guy, and if appropriate around Christmas time "Grease" your dock hands, then when you need some special help from them, they will probably be there and happy to help.

Going to Work Each Day

If you live aboard, you do not have to go anywhere. Of course this has it's downside as well. You work and sleep in the same place, like Melina Mercouri in the movie, *Never on Sunday.*

Having choices is a good part of this business. I get to choose the menu for our trips either by virtue of my going to the supermarket or telling the stew (crew) what to serve. Then usually I get to choose what time we leave and when we will return, where we go, how fast we go and by what route.

Most charterers have only general destinations in mind, or none at all; they are just "On holiday, out for a good time."

If you are running a fishing charter you will be responsible for finding the fish. Sometimes that is a problem. The fish know where to go to keep you from finding them. Of course, you have some help in knowing coordinates of wrecks, holes and ledges where they tend to congregate. Then you get some help from birds, friends with tips and eavesdropping chatter on the VHF. New developments in side scanning sonar make locating fish schools in your vicinity like, if you will pardon the expression, "Shooting fish in a barrel."

Sportfishing boats generally wear out or are retired long before a private cruising yacht ever has its first overhaul. Moreover, captains wear out long before the equipment lets go. Sometimes, if you are married....well, wives or husbands wear out even sooner. Maintain your equipment, your attitude and your personal life; they are all critical to your success.

Attitude is your greatest asset. Being old or young, tall or short, white, black or any shade between, the right attitude toward your customers, your boat and your business will make you a success.

This attitude includes being "up" all the time you are aboard or out drumming up business. Rather than worrying about all the things that can go wrong, enjoy being on your boat, on the water and invite your prospective guests and customers to share your joy.

Everyone wants to be happy, and being around happy people helps everyone's attitude.

Just putting on your logo shirt, clean shorts and presenting yourself to the world as "The Charter Boat Captain," will make you feel better than ninety percent of the other captains and your business will show it by growing.

But, please don't fall into the gold braid and scrambled eggs trap. Every now and then, some newly minted captain runs down to the uniform store and loads up with gold braid jackets, scrambled egg hats, and wanders down the dock talking loudly on his handheld VHF radio or cell phone trying to impress.

Be proud of your accomplishments but don't push it. As a captain you are one of the select few, wear your chosen uniform smartly and be on your best behavior out in public. Make us all proud of you.

Crew

Lucky you if your significant other is your crew. No one else will give you the daily support and encouragement as well as someone who shares your life. Next best is another family member.

Always take the time to train your crew so both your efforts are smooth, coordinated and safe at all times. That goes for any crew you have aboard, paid or volunteer, make sure they know what you expect of them and how to do it.

My Mother was a southern lady with impeccable manners. She welcomed everyone into her home as if they were long lost family. You could do worse to guests on your boat. Good manners, courteous behavior, respect and the ability to listen well are very necessary attributes in captain and crew alike.

When guests are present, crew always speaks to the captain as Captain (first or last name), never just by name alone. Guests have put their lives in the captain's hands for the duration of the trip, give him that respect, and help insure the guest's confidence.

I have had wonderful crew who are good with the guests, spread a beautiful lunch, and in every way are the perfect hostess, but who could also not remember how to tie a fender to the rail the same way twice. Others, could handle the boat as well as myself, but had no sense of service when it came to the guests aboard. You must keep training and fill in their weak spots. Moreover, sometimes there are things you must just overlook in crew.

Hiring and firing can be a trauma for some who have never had to make people decisions based on facts and gut feelings. You may feel pressure to hire your friends who have supported you in your venture, but are they really the best candidates for your crew.

Write down what you want for your crew, good looking, hot pants and tight T-shirts (just kidding) and see if any of your candidates actually meet the written criteria.

One piece of advice I can give from experience, is to check any references you are given. Ask questions and probe the answers. Valid, honest candidates will be complemented if you call their references; after all, they've gone to the trouble to give you good information, which they expect you to use.

Crew Pay

Ask around before you make offers. You need to meet the local pay rates, but do not upset them by paying too much when you start. After awhile if you are successful, and have found great crew and want to keep them, adjust their pay accordingly. Tips are a large part of crew satisfaction. Work hard to insure your crew is tipped or share all tips among the crew. No one should be left out. Tips are reportable as income.

Along the East coast of Florida, pay ranges from $10-$40/hour for casual labor or short time work at the dock. Expect to pay $100-200/day for experienced stewards, stewardess, or mates, more for a chef. Wages and salaries on expensive yachts and long offshore trips cost more

If you have never operated a business with employees before, go to your state employment office, your local state tax office or other local source and determine what taxes you are responsible for regarding any employees.

For anything more than casual labor (hiring someone for the day) you will need to collect and pay social security taxes, withhold income taxes, pay unemployment tax, workman's compensation tax and perhaps other "payroll" taxes that are necessary in your state. These taxes are usually called payroll burden, and while they may be a burden to your business, they are the law of the land and you must structure your business income to provide for them.

My state requires that, "A business is liable for state unemployment tax if, in the current or proceeding calendar year, the employer; (1) has paid at least $1,500 in wages in a calendar quarter; or (2) has had at least one employee for any portion of a day in 20 different calendar weeks in a year, or (3) is liable for the Federal Unemployment Tax as a result of employment in another state." You will want to verify your state's requirements.

This might be a good time to pay a visit to a tax accountant or advisor who can steer you through the rocks and shoals of payroll taxes in your state. Any practicing tax accountant will know what applies to your operation and give advice about how much payroll burden to count on, based on your estimate of crew costs.

One almost foolproof way to help yourself is to open a tax account with your bank. Then every time you write a payroll check to an employee, you also deposit a check or transfer the withholding tax amount in your tax account for the taxes and deductions you must pay quarterly to the various government agencies.

That way the money will be there when you need it. It is a very serious mistake to co-mingle your funds with those that are due or owed to the federal and state governments for taxes withheld from employee's wages.

If you have permanent employees keep good records and pay your payroll taxes on time. Conversely, a lot of boat work is by independent contractors or casual labor. Remember to keep track of your casual labor for your tax return also.

A few bucks here, a few bucks there have a way of piling up into serious money owed the government. Smooth sailing on an outgoing tide may result in a mess at the inlet mouth when it meets the IRS, excuse me, the ocean.

Other Local Licenses Required

You will probably be required to obtain a business license. This may be only a simple occupational license issued by your county or city (or both) or a fishing guide license issued by your state or other jurisdiction.

For example, if you are operating a charter boat from a city dock downtown or from a private marina in any state you will probably be required to obtain the following licenses;

- A city occupational license for the business
- A county occupational license for the business
- A state tax certificate to collect and pay sales

    taxes on your charters, merchandise, extras you sell.

- Perhaps also a food handler's certificate

Another license pitfall you may avoid if you skirt carefully around the legal statute is needing a liquor license to serve alcohol aboard. Often if you can simply give it away or ask customers to bring their own, then no license is required. Your state may have other requirements in this area, check carefully. You do not want to conflict with liquor laws. On the other hand, a federally documented vessel working offshore is not necessarily subject to state law just passing through. This may explain why many larger vessels have extensive "ships stores."

If you are a fishing charter captain, you may need special fishing or "taking" permit or licenses to work a particular area or fish a particular fishery. Sometimes a special trip report detailing the days' catch must go to both State and Federal jurisdictions on a routine basis. You may also need to report boarding and debarkations to your Dockmaster or marina office. With the growth of regulations regarding Homeland Security expect the rules to get tighter.

Routines

Unless you have some routines to keep your sales efforts propped up, most of us will fail to follow up every contact and eventually our business will fall off and then we wonder why. But there are things you can do on a regular, set basis — routine things like making sure your brochure racks are always full with your current brochure or rack card, and planning your advertising on your calendar so that you make the publications you want to be in during the time you want.

Most give-away publications, maps and vacation guides have a three to six month lead-time, do your planning ahead.

If you want to be in their fall season brochure, you must have your copy ready long before summer. Managing your advertising requires thinking far ahead and having the money available when you need it. Advertising is a tough racket, which is why some ad execs make so much out of it.

Routinely visit your Chamber of Commerce office and chat with the Information Specialist or counter person on duty. They are the person most visitors will see if they stop by or call for information on what to do in the area.

If you have been in their office lately with news about all the fun your charterers are having or the latest fish story and a big smile on your face, they will remember you and recommend you to visitors. Many times I have people walk down the dock and comment to me that those, "People in the Chamber office over there recommended you." When that happens, I always stop by the next day and thank them personally.

If you get business from hotel or motels in your area, the same holds true for them. Pay them a visit on a regular basis, get to know the desk clerks by name, be interested in their work, commiserate with them and if you have a chance, invite them for a ride on your boat. They will remember you forever and send you business.

Another type of routine is maintenance on your vessel. I probably do not need to tell you to watch your oil change intervals, but if you develop a routine of doing all your maintenance items at a particular time or day, you are much less likely to forget something important.

In Florida, we run air conditioning aboard at least half of the year, more to control the humidity than anything else. Unless you are in an area with perfectly clear water, the intake filter (you have one I hope) for the AC cooling water may clog up with crud in a very few days. Sometimes I have to clean our filter twice a week. Other times it will go a couple of weeks, but it must be cleaned or the system will not work properly or not work at all. You probably have other necessary routines to follow.

Another must-do is draining the water off the bottom of your fuel tanks. If you are fortunate, there is a tap on the bottom of the tank and room to get to it. Once a month open that tap over a container, drain off any water, and sludge in the tank. Then do the same for the bowl of your primary fuel filter. (The primary filter is the big one between the tank and the engine; the secondary is the filter on the engine itself.) By keeping the water out of your fuel, you inhibit growth of algae and the collection of crud in the fuel tank itself.

My engine also uses the fuel to cool the injectors and as a result passes a lot of fuel through the filters and back to the tanks. In a good day's run, I will pass all the fuel in the tanks through the filters at least once, helping to keep the fuel "polished" and clean. Some diesels are not so engineered or the fuel tanks are so large that the entire fuel load is seldom is passed through the filter.

These are the boats you see having their fuel filtered and polished dockside and whose owners are always buying fuel conditioners. If you can drain the tanks regularly and change your fuel filters often, you will avoid many engine fuel problems.

If your boat has built-in tanks that have no bottom fuel tap, try to rig a suction pump to slurp up that bottom stuff through the filler cap or wherever you have an opening or cleanout. If sludge stays in your tanks, you will pay. I might also mention that running your engine(s) for 10 minutes at the dock once every couple of weeks will not, repeat not keep you from having cooling, lubricating and fuel supply problems with your engine.

Engines need to be run frequently, no less than once weekly for an hour or more under load. The reason for that time is an engine needs to get up to operating temperature and stay there long enough to vaporize and blow off any condensation moisture in the crankcase. It is condensation water in the oil that ruins your bearings and rings. Diesel engines in boats are rarely worn out from overuse, but they will often fail from just the opposite.

Another housekeeping routine I scrupulously follow is cleaning up any food crumbs or open packages of food—ants and roaches will find them. Once you get either of them aboard you will probably have them for life unless you lay the boat up for a year or more. Combat® makes roach traps and ant traps that work, but they are effective only for a short time after being opened and the traps must be changed every three months or less. For bad ant infestations, try sprinkling fire ant killer around on your decks. Leave it for a day or two if you can, then wash it all off.

A trick we used in the Caribbean if we were going to be off the boat for awhile was buy a pint of formaldehyde from the druggist (God only knows what they used it for) and pour it into shallow pans low down in the boat. As it evaporated the vapors killed roaches, ants, mildew, and anything else that might be aboard if you left the boat closed for a few days.

Scrubbing the bottom gently will re-activate your antifouling bottom paint. It keeps barnacles and other nasty's from obtaining a foothold aboard. Scrub too much and you brush off all the paint, but a gentle brush cleaning every few months or so will keep your antifouling active.

Find a spot of clear water where you can go overboard and do it yourself. You will learn where the antifouling wears off first, can check for any prop damage, and clean all your water intakes.

If you hire crew, give them lists of the routines you expect them to accomplish. Everyone feels better knowing what is expected and you get a better overall charter operation when everyone does his or her part. I am not suggesting micromanaging your crew, however, if you want certain things done every day, put them on a list for your crew to follow. Then there can be no confusion or doubt.

For me, some of the routines that give me pleasure are handling my vessel in a seamanlike manner when docking or undocking. Every time I make a landing I seem to learn something. Showing the flag from 0800 to sundown is a great way to show your pride in your vessel. However, never leave it flying after dark unless you spotlight it. Since I don't live aboard I put mine out when I come aboard and take it in when I leave. It tells everyone who knows me that the captain is aboard. I have a canvas sleeve that goes over my flag to protect it from the sun when I am not aboard so I can leave it out in its socket, ready for flying.

On the bow, I fly the Union Jack on Sundays and holidays. It is a beautiful flag consisting of just the blue field with 50 stars of the U.S. Flag. Other times I fly a private pennant at the bow staff. My current one is a swallowtail flag with the year my boat was launched, 1936.

On special occasions like holidays, festival weekends and when we perform weddings aboard, I fly a string of signal flags from the bow across the top of the mast and to the stern. Tradition dictates a weight on each end of the string of flags so that the flags pass over the bow, stern and reach the water, but not in it. You can spell out any message you want and perhaps some salty person will be able to read it.

I spelled my name once in a run of flags we hung in a bar for decoration. Some months later, while sitting at that bar, I was startled to hear some old salt bellow out to no one in particular, "Who the hell is Conrad Brown?" I bought the man a drink!

Another perverse routine I enjoy is polishing brass in the wheelhouse of my antique yacht. I don't have to do it often, but every once in a while, making sure the brass shines is a special treat for me. I once had a stewardess aboard for a couple of months who kept the brass polished to perfection. I rather begrudged her doing that; I felt it was my job even though she did a better job at it than I.

Keeping the logbook up to date is a very worthwhile routine, but one I find hard to do. A daily log will help you develop your routines and jog your memory when it begins to fail, as mine seems to be doing. My guest log in the wheelhouse gets a lot of attention by guests and they always have a great interest in who has been aboard and where they are from.

Always encourage your guests to sign and comment in the guest log. It is a great source of advertising ideas and helps you remember memorable guests and events.

Office Expenses

Here are some expensive thoughts on having an office or hiring someone to represent your business. Overhead, rent, all these things accrue when you do not do it yourself.

If you possibly can, eliminate all fixed expense such as dock rent, landline phone, fax line, phone book advertising, unless it is absolutely necessary. During the lean months, and there will be lean, slow months anywhere, holding your fixed expense to the minimum will make you sleep better at night knowing your business is not hemorrhaging away your hard earned money. You can worry about the big expenses when you need to but do not ignore the little expenses that month after month eat away your profit. Check your cell phone bill, watch out for add on fees at your bank or credit card-processing company. Pay only what you must, conserve your cash.

Some locations will require an office, either to sell tickets for your operation or if you are out on charter, or out of the area for the season and out of contact for any period. When chartering first started, there were charter clearinghouses whose main purpose was to keep the calendar of charter boats so that brokers could call and see if a particular yacht was booked for a particular week, etc.

This is still the case for large luxury, expensive charters where it is broker talking to broker, arranging for a client. Several large "houses" handle certain yachts exclusively. You are probably not there yet but the time may come when you can establish yourself, your boat or both with a brokerage house.

For now, your business will probably operate with a cell phone in your pocket and a calendar in your hand while you are getting started.

Steer your business away from making charitable donations the first three years unless you can see a direct benefit to your business. Smile and say, "I'm sorry; we are a new business starting up and must conserve our operating funds for the business, just now."

Ask them to come back next year; possibly you can help them then. You will have made them feel good about your good intentions and you will probably see them again next year. Maybe you can make a sale to them or their friends then. Be sure to give them several brochures for their office as they leave, and thank them for thinking of you. Make a friend wherever you go.

There have always been charter vessels, beginning with ancient Greek merchants chartering ships to carry their merchandise and raw materials. They didn't want to own the ship, just transport cargo.

From that beginning, the procedures evolved to chartering a boat for hauling cargo and later chartering just for pleasure. In the early years of yachting, it was common for wealthy individuals to charter a yacht for a holiday or to participate in some yachting activity like race week or the whole 'season' at Newport or some other wealthy watering hole.

The small boat charter business started in the U.S. Virgin Islands in the late '40s when Basil Symonette sailed some tourists around the island of St. John.

By the early 1950s there were several boats offering a week aboard for about $900 for the yacht plus $6 a day per person for food and drink.

Back then Mt. Gay rum was eighty cents a fifth (sixty-five cents if you brought your own bottle to be filled at the Rhum works on Martinique) and Coke® was about $24 a case, making the mixer rather more expensive than the rum. A lot of rum and water went down then, and if you knew of a lime tree somewhere you could get too by dinghy, well….life was sweet.

In the late 1950s, Sparkman & Stephens and a few other brokers began helping clients and boat owners get together on the east coast. By the early 1960s, Nicholson's at Nelson Dockyard in Antigua were placing charters on large, luxurious ocean going sailing boats cruising down island.

Now charter brokers and charter clients deal directly with captains or business agents via cell or satellite phone and get confirmations quickly. Clients book charters all over the world, as many have been chartering so long they are continually searching for new areas to explore by water.

This direct calling puts more strain on the captains but also puts them more in control. My business runs with a cell phone in my pocket and a Day-Timer® in my hand, which keeps my calendar of bookings, expenses, credit card charge slips, address list, and phone numbers of suppliers. The KISS principal works best. Remember, we are here to have fun.

Chapter 6

Communications

"The public is like a piano, you just have to know
what keys to poke."   Al Capp

Marketing 101

Success in your business depends on communicating with your clients effectively, honestly, and timely. You must exude confidence from the first word you speak or print, confidence that you can satisfy your customer's needs if they purchase your services. This applies whether captaining or chartering.

I answer my phone with *"River Tours, this is Captain Conrad, how can I help you?"* This greeting is informal, informative, and friendly. From my first word, clients know they have reached the right company, the right person and that their questions will be answered.

Employees or family members answering your phone with just "Hello" or mumbling the name of your business is not the way to greet you clients.

Another thing, be sure to smile when you answer and say your greeting. The smile comes across in your voice, it sounds just like you were waiting for 'their' call.

You Are Your Business

This sounds simple enough, but it takes some getting used to. It means that all day, every day, everywhere you go; you are selling your service, your business, your way of life. You are your best advertisement. This means you will always look professional, presentable and clean. No ragged shorts soiled T-shirts or dirty sneakers.

OK, how about when you are working on the boat? To reinforce my professional image, I have khaki work clothes (with the boat name on them) to wear when I'm into some big work project on the boat. They can get dirty without looking dingy like dirty whites and I save the wear and tear on good knit shirts and shorts. However, do not parade around in work clothes, shower and change as soon as you can.

On our charter river tour, my crew and I wear white shirts with epaulets (e-pa-'let) and our names embroidered above the pocket. Guests sometimes have a hard time remembering the crew, so names on the shirt give them a little help.

For my everyday uniform while not on charter, I buy a good quality white knit shirt and have my boat name embroidered by a local sports embroidery store. You might want a distinctive color, maybe to match your boat's color and coordinate all your advertising, crew uniforms or tee shirts. Sell your shirts and caps to customers for extra income.

Epaulets are shoulder boards signifying rank. Four gold bars are a captain, three for Executive officer or second in command while two is for department heads like engineering or mates. One bar is for the lowest ships deck officer.

Stewards, stewardess, and chefs use the same designation except using silver bars. Someone in a strictly inside position (housekeeping) has a quarter moon on the shoulder board sporting silver bars. These are not hard and fast rules, you will see many variation including waiters in dockside restaurants sporting four gold bars....making all real captains cringe.

Having lots of red, blue or some other distinctive color walking around advertising your business could be a good thing. Whether going to the hardware store, marine office, or just going to lunch, I always have a few rack cards in my pocket. When I pay my bill, if it's to someone new, I pass along a card and invite them to come see the yacht.

> *"Lunch was great. Here is what we do,*
> *stop by and see our boat sometime."*

Business Cards

Consider this about business cards. They are important when you want a small card to give someone with your name and how to reach you.

Make sure it has a clear picture of your boat also. But for prospective customers, give them a rack card. There is much more information on it and, being larger, it is much less likely to be lost.

People love to know what you do. They see your shirt and by your bearing know you are special. Giving them a rack card gives them an opening to ask a question, which in turn gives you a chance to chat with the person and tell your story. (Remember your one sentence description and work it into the conversation).

Practically everyone knows someone who would like to go out on your boat. They will carry your brochure around and talk with friends. I've had people call months after picking up a rack card. They were just waiting for the right time to go on a cruise.

The trick is getting the word out everywhere to anyone with any interest in boats or the water. Many people secretly dream of being a charter boat captain. I have spent many charters explaining how to do what I do. This book is an outgrowth of those conversations.

Innumerable surveys have shown people want to be their own boss, manage their own career and have a wonderful lifestyle. People will pick up your information and carry it with them for perhaps years before they are ready to take that first step, namely, go on your boat.

Communication involves web pages, brochures, rack cards, business cards, flyers, signs and listings in trade journals, print ads and your boats' appearance, your attire on the docks and around town, all these things are communicating your message. Make every effort to tell your message well.

Personal contact is the cheapest form of advertising and the most effective. Get out and meet people, tell them what you do.

In all your advertising, have a consistent message, one that drives home your message point by point. Pick a theme for your business and let everything reinforce that theme. Review your 20-word description and your one sentence description for theme ideas.

Marketing Themes
- Fishing charters-show the big fish, happy angler, smiling captain.

- Luxury term charters-great dinners, happy guests, and luxurious surroundings.

- Ecotourism-jungle environment, wild animals, whales and dolphins jumping.

- Tour boats-comfortable seating, big windows, shade, knowledgeable guides.

- Whale or bird watching-comfortable seating, open decks, binoculars, cameras, guide with book and binoculars in hand.

- Sailing charters-Rail down sailing, spray, guest at the wheel, smiling captain.

You get the picture. Whether these word pictures require artwork, photos or a combination of the two, you must communicate effectively in pictures what you do. Pictures are worth a thousand words.

Carry your theme throughout your communications with your clients.

Advertising

Follow Ted Turner's advice, "Advertise, advertise, advertise and work like hell." How to begin? Study your market. Who are the people you are trying to reach; young professionals, the occasional angler, families, or seniors?

Each of your target customers will respond differently to advertising. You may need different kinds of ad material or media to reach each of these different markets.

If you are offering fishing charters and are one of many charter boats along a particular dock, your dock sign or 'sandwich board' may be most important. But if you can snare customers before they ever get to the dock, you've eliminated most of your competition.

How do you do that? You go after your customers where they are, in their homes and in their offices. Put your message directly in their hands, so to speak. How? Reach them on the Internet.

Internet Web Sites

Since you logged on yesterday, the Internet has changed. Face it, the Internet changes second by second however, like our fancy new GPS or chart plotter; we can learn to use the Internet to make our business better. Do not be afraid, just plunge in, and get a handle on it.

The Internet is my best and most effective single form of advertising. It is nearly free; and I reach practically everyone looking for a charter boat or bed and breakfast in my area. I change the copy or prices anytime I feel they need updating and can add or delete trips or special offers easily.

I generally work on some of the web pages weekly, updating information, adding or changing photographs, anything to make the pages more friendly, usable, and fresh.

Always ask you clients where or how they found you and what in particular caught their eye. They are happy to help you fine-tune your advertising; it makes them feel responsible for your success.

In spite of all the Internet activity, I do no bookings directly through the Internet. Instead I ask people to call so I can get a feel for them, discuss any alternates if the dates they want are booked and get a credit card number over the phone to confirm their reservation. It gives the customer a better feeling if they speak directly to the captain, knowing they have reached someone they can trust, who will not misuse their credit card or other information.

Having their credit card number creates a moral contract between the two of you and helps insure they will show up when agreed. If you have a clear policy stated on your website regarding cancellations, it helps cut down on "no shows" also.

Building up your web pages is not easy.

If you are not computer literate enough (yet) to use FrontPage®, Publisher® or some other web page program to design your own web pages, go on the Internet and buy a web page already formatted. (Search Google for web page designers.)

You may still need some help to get started. Your local computer store or vocational school can offer web page design help, or can point you to someone who can. Once your pages are up and running, it is not a big deal to update and/or change them.

You can always hire someone to manage your web site, but be sure you are happy with the result. I get calls all the time from someone wanting to re-design my website free just for the experience. The catch is that then I have to use their web hosting service for $$$ a month.

That also allows them to lift my e-mail address and sell it if they are so inclined. The Internet is a vast sea of opportunity, with all manner of fish to catch in your nets, but be careful of the sharks, rocks, and shoals.

Every location has different web hosting facilities so you should scout out who is providing the best service in your area and talk to them. Or, go national and look for the best deal on the Internet.

Do a search for -web-site hosting- on any Internet search engine. Find one that fits your budget. I pay just twelve dollars a month on an annual basis to a local provider to host my web pages. Service could not be better, and they are always there to answer questions and help.

Your provider does not have to be local, but I like to keep my money working at home and if there is a problem or question, the answer is a local phone call away.

Some major Internet providers offer many extra services for very little money. They may be best for you, ask around from others who have web pages, your Chamber of Commerce for instance.

Internet Names

You will need a name for your Internet site. Go back to the ideas for your business name. Think again how you can integrate your business name and your web site address.

For example, my business, Old Florida River Tours is too long for a web address. If I tried to use it would look like this:

www.OLDFLORIDARIVERTOURS.com

That is much too long and complicated for anyone to remember much less write out correctly. But www.OFRT.com is short, unique and easy to remember. OFRT, (Old Florida River Tours) of if you prefer, Old Fart, it may not be pretty, but everyone remembers it.

My business depends on the web for at least half my calls. However, at first, when I posted my web site, nothing much happened. None of the search engines could find me because my pages were not set up properly for them to find.

At this writing there are 2,560,000 listings on GOOGLE for Charter Yachts.

You can begin to see the difficulty in being listed in the first 10 or 20 positions. But pick the right name, optimize the site design for the phrase and promote it well and you will land in the top ten. If not, and it is important to your business, you can purchase access to the top listings.

Search engines look for specific data when 'spidering' a web site. There are six areas, some of them visible on the home page and some of them located within the HTML code, and therefore not visible.

These can be optimized with very little change to the existing web site. The data that robots look for is contained within the following six areas on your web site.

- The Title Tag (visible)
- The HI header (visible)
- The Description (not visible)
- The Body Text (visible)
- The Alt Tags (visible)
- The Text Links (Hyperlinks) (visible)

No, I do not know what all these are exactly either, but fortunately there are web designers who do and they can help you. Find a web site designer at your local high school, computer store or Google the web for web site designers or optimizers.

The site, www.WordTracker.com can help you identify the most popular words about chartering, fishing, or whatever.

Once you identify the right words search engines are looking for, use them on the first page of your web site. Periodically review your first page to be sure you are using the current words search engines are looking for.

Avoid Frames in the design and Flash pages, as these are difficult for search engines to index. Stay away from sites where you piggyback on some other site, even if they are free. For example: (www.chamberofcommerce/yourbusiness.com) Search engines will blast right by your site with nary a glance.

Have at least three pages on your website, more is better. Many search engines (notably Google) like informational sites with lots of detailed information. If you can link your site to other similar sites then search engines will turn up your site while searching for others. You can check what search engines want by clicking on:

www.SearchEngineWatch.com

A couple of weeks after tweaking my web pages, I began to get 'hits' from people searching for yacht charters, slowly at first then more and more.

One aspect of the web is that the more people find you, the higher you rank in the search engine hierarchy. It's like the more hits you get, the more hits you get. Weird.

Another tip is; <u>lose the prose and present the information</u>. People looking for information don't want to wade through your personal attempt at writing English Literature to find:

Who, What, Where, When and How Much.

If you are good at writing and have a lot to say about yourself and what you offer, put it on another page. You can call it about "About Us" or something like that. Some people really like to dig into stuff, your history, your boat's history, etc. Keep your landing page simple, with a couple of quick loading pictures and enough copy to intrigue your customers to stay and look around your website. Remember that fifty-word exercise, use it here.

Set up separate pages describing your area, one page for the fish you catch, another for the whales you watch, the ports you cruise to; all the details your charterers can expect. Clear maps are good. Net surfers love details and photos that load quickly.

Link all these pages to your home page and put a hit counter on each page, hidden if you can manage it. The hit counter will tell you how many people are searching your site and what pages they are looking at, prompting you to beef up those pages that are not selling effectively. Track if you are getting more hits this month than last. If not, figure out why and change something. Get free site tracking links in your website. Search for "website tracking programs" on Google.

Print your website address in your brochure or rack card. You can change or modify your trips, tours, charter, etc. any time you please; no more big print bills. Likewise, post your prices on your web page, then as you need to change them, it is easily done and everyone has the same information at the same time, clients, brokers, media, etc.

Many links to other related sites will get you boosted in the search engine rankings. The more related links you have will work to your advantage when people are trying to find you. Link your site to weather sites, Chamber of Commerce, yacht clubs, professional marine related groups and other charter boats. These links should be reciprocal links if possible; they generate more exposure for you.

Currently, links back to your website from other websites are a major factor in Google's ranking algorithm. In addition, link words in your home page to other pages on your site. This allows search engines to spider all your pages, not just the opening page-more hits.

Scan in lots of pictures of what you do. Make them no larger than 640 pixels maximum width, 300 is better. You want your text and pictures to fit within a small screen without scrolling. Shrink pictures and text to leave some 'white space' around, which make your pages easier to read. Use 'thumbnails' that expand when clicked. Above all, your web pages should load quickly so people will not give up and go looking elsewhere while waiting for your information to load.

I personally do not like sound and music coming from web pages, it takes longer to load and the distraction of someone else's choice of words or music is annoying.

I have found the following guidelines work best for designing websites. You will find more if you look.

Web Page Guidelines

- Less is more.

- Use a dramatic image.

- Be #1 at something and say so.

- Storyboard your site for interest and clarity.

- The first page should load very fast.

- Ask you customers for suggestions.

E-mail addresses are usually available with your website and are good to use exclusively for your business, keep your personal e-mail separate. You may need to set up a spam filter to keep the junk out of your e-mail box. Scrolling daily through a hundred offers for physical enhancements may be fun, but you have better things to do with your computer time and besides, you may inadvertently delete some important mail buried in all that spam.

I'm sure my web pages could be better, but by doing them I can fix or add anything I want, whenever I want. The feedback

I get is that my site is friendly, informative and pleasant to view. Make your site have your personal touch, a professional approach or fun, attractive thing to do. It is your first chance to project yourself and your business to a prospect.

Rack Cards

Unless yours is the only boat available to cross a wide and dangerous river at the end of a long road in the jungle, you need to advertise.

Plan to spend money up front for brochures, rack cards, business cards, flyers, signs, billboards or skywriting, whatever works for your market.

Your target customers must know you are there, offering what they want and how to reach you or you will go hungry. In other words, advertising pays.

A good rack card (coated card stock 4" x 9" both sides printed in color) is basic to your business. Customers will pick them up, carry them around, sometimes for years, and expect to get what you offered when they call. (Rack cards are better and cheaper than multi-fold brochures, they communicate faster.)

Rack cards will say or, better yet show what you do with a picture near the top of the card so it will show up well in a rack full of other, competing cards.

Then your identity, your business name, your name or your boat name and where you cruise or fish, etc.

Your boat name is of equal importance to the description of your business. You must distinguish yourself from whatever competition you have. Be creative here with names, but don't get 'cute' or ridiculous with your boat or business name. Unless you are running a party barge, you want to project competence, proficiency and expertise at what you do.

Include some good pictures, with captions or copy showing what the customer can expect on your boat.

Next, in BOLD type, how to reach you; phone, location, slip number, etc.

List your website (www.MYBOAT.com) on the front also; it is as important as your phone number.

That's it for the front side of your rack card. Good, solid basic information, but please no life stories here. You really want them to pick up the phone and call you, the sooner the better. Over the phone, you establish personal contact with the potential client and close the sale, or at least get them to follow up on their phone call and start planning for their trip on your boat. You can then follow up in the most appropriate way, mail, e-mail, phone call or personal visit.

On the back or other side of your rack card, give details. A 20-word description of what you do. Stay away from wordy type on your card. Make it brief, snappy and to the point. Refine and re-define your twenty words.

Briefly tell what you offer, when, where, what to bring. Leave pricing off the rack card. I've tried it both ways and it is difficult to adjust your pricing to meet the demand or competitors if your prices are printed and widely distributed. Later if you feel you must add prices to your brochures, print stickers to put on the cards. You can print these stickers on your home computer using Avery templates; almost any office supply store has them.

Show more pictures of what they can expect, talk up the trip briefly and give your phone, location and website again in large, bold type.

If you have room, throw in some good quotes from previous customers.

Make the type as large as you can for easy reading, some of your customers will appreciate having larger type.

Get people of different ages to read your "dummy" rack card, see what they think. Do they "get it?" If not change it.

Obviously you have picked over all the rack cards in your area looking for examples that grab your attention. How does yours stand out among them?

Imitation is the sincerest form of competition here.

Don't steal their copy but look for ideas you can use. Do what they do only do it better!

Print rack cards in color on both sides. Nothing sells like color and the added expense, if any, is fully worth the money. Have your printer use stiff card stock, coated to accept color printing well. Ask around the printers in your area for their recommended number of rack cards to print, ask your competitors, Chamber of Commerce and others in the rack card business. They will steer you right. Have your printer save the printing plates for your card, and then when you need more later, you only pay for paper and setup.

Yellow Pages

Yellow page advertising is expensive and not nearly as important as it was before the Internet. If you have a fixed base and an office or if you sell tickets to your business, you may find Yellow Page advertising to be useful. Be mindful however if others are advertising there, you are directing your potential customers to your competitors as well.

It is easy to fall into the trap the attorneys have succumbed to, each one trying to outspend the other buying yellow page space. Yellow page contracts lock you into a twelve-month contract, and whether your business takes off or not, you will be buying yellow paper for twelve months. Only the yellow page directories win that game.

Newspapers

Most charter yachts avoid newspaper advertising for the same reasons. General circulation newspapers are too expensive for most charter businesses. Focus your advertising to that 3-5% of the population that is interested in what you offer. Do not waste money preaching to the other 95%.

Local dockside papers may reach your market and the customers trying to find you. These papers might warrant your attention. On the other hand, tour boats that run a daily schedule may need newspaper advertising to keep their public informed.

Radio & TV

Avoid Radio and TV spots, unless the spots reach your market for a special event. Avoid magazines unless they reach your market directly. Magazines have such a long lead-time; your business may have changed by the time your ad reaches paper and ink. Upscale 'things to do' magazines for a particular area might work.   Travel magazines can be good and neighborhood newspapers are effective if your demographic customers live there. However, have an overall marketing plan. Do not shoot advertising dollars out with the buckshot approach. Be selective, pointed, and effective.

Track the business you get from every form of advertising you do. If you can't track it, it's not working. You will quickly find where your dollars are most effective.

Your message must always be clear, concise, consistent and timely and reach your potential customers in time for them to make the decision to call you.

Get professional artwork for your brochures, rack cards, business cards and any advertising that appears in print. Your advertising must be consistent and similar through all media to help capture your clients through repetitive recognition. Develop a logo featuring your vessel or activity in a stylized form or settle on great action photograph and make sure it appears on everything you print and in your web pages.

Free Publicity

The 'finest kind' of print advertising is a news story in a local newspaper or magazine. If you can arrange for a reporter or freelance writer to do a story on you and your boat, you will see immediate results.

Customers will seek you out carrying the article in their hand just to see if it really is as the story says. People have great confidence in what they read on the printed page. Once they've read the article about you, they want to get on board. All you have to do is deliver.

I still get calls from an article about our area that ran in the Miami Herald months ago. In the sidebar of places to stay was a short blurb about staying aboard a classic wood yacht and taking a cruise for a special price. Every now and then someone calls about the "Miami Herald stay & cruise." Wonderful!

An author friend called one day and asked if he could write a story about us and try to sell it to some local papers? Of course, I said. We spent a pleasant hour together; he snapped some photos, wrote the story and a local paper picked it up. He made some money and our boat got wonderful free publicity.

Another author friend brought her family along for a short cruise and then sold that story as part of a series about working people to another paper. Her article gave us more great publicity. Suggest these ideas to reporters and authors you meet. Who knows? Free publicity is the finest kind.

Many TV stations use boats as a backdrop for marine related stories or they may suddenly need a boat to transport a film or TV crew to a marine scene, disaster or otherwise. Be helpful, do them a favor. You can provide some marine background material; you are after all, a professional captain with your own business on the water.

They may not work in a word about your boat or yourself, but just being friendly and available to the station personnel will get you mentioned from time to time. Your boat in the background of their story doesn't hurt either.

A Boston TV station was in town for a Red Sox game and interviewed a player who liked to fish in his free time. The interview was conducted on our foredeck. The camera swept across our boat, then across our name board to the player and the short TV interview clip appeared on the Boston evening news.

The next morning I got a reservation call from a viewer in Boston coming down for a game in a couple of weeks. He loved the boat and after staying aboard for two nights, brought some former-Red Sox Hall of Fame players aboard for a short cruise. All from a few seconds free exposure on TV.

Some boats have found a part time person to generate copy and press releases for local media consumption. If you can pay them a percentage of new business they generate, your business will not feel their cost and the new business they bring in will improve your morale and bottom line.

Some will work for just a boat ride from time to time. Trading is nearly always an option when buying some local service, particularly from another small business, your printer, the T-shirt shop, or your bait shop.

Find out when you are likely to see dock strollers or potential customers on your docks and be there when they are. Weekends, holidays are good, festivals are better and dock parties the best. Try to arrange some event near your dock and be there in uniform. Talk to everyone. Pass out your rack cards and invite likely prospects to come aboard your boat for a look around. Naturally, your vessel will look good, clean and ready for guests with everything not needed stowed and out of the way. You will develop your own techniques for determining the valuable prospects from the merely curious.

Many of my customers were introduced to the boat from a quick tour during a weekend festival or other activity on the docks. They would take my brochure or rack card and dream, eventually making the call and taking a trip.

One cautionary word, avoid bad publicity. It seems obvious, but bad publicity about any other boating activity will wash over you as well.

A charter boat named POSEIDON had to change its name overnight when the disaster movie of the same name came out. Brokers were calling the next day to cancel charters already booked. (Captain Walker responded to the challenge by having the brokers tell their clients they were being shifted to a newer boat named CARRICK) It was of course the same boat, with a new name. Ergo, you must be positive and always in charge of your image.

Pick the name of your vessel carefully as well. Make it reinforce your image but be sure you have the right image in mind. The charter fishing boat, Monkey Business will forever be associated with Gary Hart, a presidential candidate wannabe whose run for office evaporated after one trip to the Bahamas on the ill-named vessel (with someone not his wife).

Recently a marine park tour boat sank in shallow water nearby. No one was hurt and everyone made it safely to shore. But it brought out fears, which affected many of my clients. For a while, nearly everyone coming aboard asked me if I knew of the sinking and what did I think. What they were really asking was, "Are we going to sink?" My reply assured them that they were safe with me.

" No ma'am, not while I'm your captain, enjoy the boat ride."

Trade Associations

As soon as you are able, secure a membership in local business associations like the Chamber of Commerce, local tourism organizations, state tourist associations and any group that targets your potential customers. Listings or links in Chamber of Commerce web sites are one of the most effective means of Internet advertising. Visitors often start at Chamber of Commerce websites looking for information or recommendations. If you are there with a listing or link, they may go no further than directly to your website.

For a local business, a Chamber of Commerce listing will be one of your best sources for new business. Be cautious about spending money to appear in Chamber magazines or other promotional flyers. It's better to put your flyer directly in the customer's hands, not in the Chamber of Commerce publications.

I am a member of the local Chamber of Commerce whose offices just happen to be conveniently across the street from my dock. I have a link on their website directly to mine and have access to all the other local business members via newsletters, flyers, visual exposure at lunch meetings, workshops and seminars. The members recognize me because I show up wearing whites with shoulder boards. (The women members love it!)

The chamber holds a monthly lunch meeting and when I show up dressed in whites, with gold bars, and with one or two stewardesses on my arm in white shorts, the room takes a collective deep breath!

As a result, these members are all aware of and help promote my business. They want me to succeed so I can use their services and I do, whenever I can. My printer, shirt embroidery supplier, a marine store and SCUBA store all get some business from me and they in turn promote me with card racks prominently by their cash registers.

Our Chamber of Commerce is very much a scratch my back I'll scratch yours type of organization where members offer discounts to other members and the savings in what I buy from them often equals the cost of my annual Chamber dues.

The other local organization I use is the "Attractions Association." It is very similar to C of C but focused on hotels, bus tours and distant travel agents bringing guests to our area. The Association publishes a guide and makes sure every concierge in hotels and motels has up to date brochures and rack cards of member attractions. You probably have some trade organization or group of like-minded business where you are. Seek them out and join if you can, just don't spend a lot of money.

At the state level, we have one of the best tourist associations, FLA-USA, www.flausa.com. For a small annual fee, they email me a list of everyone looking for information to write about or photograph new, unusual, special, romantic, adventurous places in the state.

I follow up with emails, brochures and invitations to writers to visit, stay overnight (complimentary of course) and give any visiting writer/photographer plenty to report on.

This has resulted in our free listing in several guidebooks with photos and nice write-ups.

There are several trade associations relating to captain's issues, charter, and fishing and ferryboat concerns. However, unless you can see a direct benefit to your operation, wait until you are well established before paying any dues to any captain's professional organization. Conserve your money you will need it.

In the Virgin Islands shortly after the charter industry became a viable industry the charter captains formed a loose organization, the V.I. Charterboat League, www.vicl.org. Their get-togethers are a great place to meet, swap stories, learn from each other and run informal contests for the best food recipes.

These types of professional organizations are an important link between you, a member of an industry, and government. Often it is this type of organization that is the only voice of reason on local issues affecting it's members and the general public regarding moorings, anchoring fees, water wildlife, and tourist taxes for water related recreation.

The Manatee issue in Florida is one of those for example. Members of some local marine trade organizations (and Boat US) are among the few reasonable and informed participants in the issue.

Participation in these local organizations keep you involved in local issues and help create a sense of belonging to the greater community on the water. Often we are so wrapped up in chasing customers, working on our vessels, and being out on the water that we seldom see anyone else.

Get to know your fellow captains on a personal level. Help them out when you can.

I know many cases where one charter captain stepped in without hesitation to help another when needed. I was out on charter once in the Virgin Islands when my son was in a car accident at college. Another captain heard the message on the VHF (my radio was off); he found me and stepped aboard my boat to finish the next three days of my charter with my guests while I flew home.

Other times captains have lent a hand to repair a generator, deliver supplies, messages, guests, and help any way they can.

In our marina, charter boat captains help each other out all the time, from manhandling a new generator installation to sharing information about waterway hazards. Often during our threatening storms during hurricane season, the charter boat captains are the ones out there getting wet, securing everybody's boats. They're professionals and show it.

Photos & Testimonials

Nothing gives your potential customers more confidence in you and your boat than photos of other customers, having fun on your boat. Incorporate photos from your past charters on your web pages; put them in your printed advertising and aboard the boat in photo scrapbooks. Be sure to get a signed release from anyone who can be recognized in your pictures before you publish his or her photo, particularly on the Internet.

When disposable cameras are on sale at my local discount store, I buy several and keep one or two aboard to snap photos of guests who forget to bring a camera along. Sometimes I develop the pictures and send them to the guests; but usually I just give them the camera. Either way the guests are thrilled.

The other use for a camera is to photograph rogue boats throwing up huge wakes or doing other stupid things that may result in damage to my boat or guests, or acts that might wind up in court. When boats throwing huge wakes do not respond to calls on the VHF, stepping out on the bridge with a camera in hand sometimes gets their attention. Of course, some believe 'they' own the waterway and nothing gets their attention. They earn a special place on the jerks bulletin board and I have their picture to prove it.

Being the Expert

If you find a special public relations person to help make sure your business is noticed, he/she will probably get you quoted in print from time to time for some waterway or boating issue.

Suddenly, you become the "expert" on water issues. You are after all, the professional captain of a business on the water. Who better than you to discuss water related issues? These issues might concern manatees, dolphin, wading birds, high tides, low tides or anything else in the marine environment. Think now of some short, pithy comments you can roll off your tongue without hesitation about local issues.

- Manatees have a right to use the waterway too, just like us.
- We need sensible government programs, not oppressive regulations.
- Working on the water is a splendid profession and an honorable career.

- People are naturally drawn to the waterfront; help us keep it always so.
- Educated boaters are the best defense against accidents.

You will have your own 'hot' buttons to push, but just make them palatable to the general public and to the little girl in Iowa who sends her pennies to "Save the Manatees." Put your own positive spin on sensitive issues. You want to inform not make people angry. In particular, you are a good source to comment on tourism issues regarding the waterfront. The media is hungry for quotes, go scatter some bait...

Make a special effort to accommodate reporters, writers, travel agents, brokers and others who make their living from or writing about the waterways. You will always receive good value for any time spent talking to people about what you love.

If they are interested in you and what you do, they will tell your story in amazing places, you just never know when something will come back to you. I have been surprised more than once with articles written for far away papers or magazines from someone who only stopped by and casually asked a few questions. And I will admit it is a thrill when the person on the other end of the phone says, "This is the New York Times edit desk, could I check a few facts?"

Whenever you change anything about your vessel or your business, it has potential as a "news story." You add a new route to your cruise, your license is upgraded by the Coast Guard, you hire new employee(s), buy a new vessel, offer a new service (weddings, ash scatterings, dinner parties, bed & breakfast, birthday parties, kids cruises, etc.)

This information is distributed by press releases, publicity releases, newsletters, and e-mail to past and potential customers and by direct mail. Your release should be news, not "stuff or fluff." Tell something no one else knows, share your insights with others and include quotes of real people to tell your story. Always include photos, even if only of yourself. People want to put a face to facts, make sure they see your face there, Captain.

If by some good fortune you get some famous or well-known people aboard, ask if you can use their photo in your advertising. Often they want the publicity as much as you do and if it is tastefully done, will be happy to oblige. Even if they only sign your logbook, try to get a photo of them on board, with you in the picture if possible. Keep this in your logbook for guests.

I keep a separate handsome logbook for guests aboard. They love to look through and see who has been aboard and where everyone is from. I encourage them to write comments in the log about their trip, the fish they caught or someone's birthday or wedding aboard.

When all else fails and you need some publicity, hold an open house aboard with some burgers & hot dogs, soft drinks and snacks or Champagne & Russian fish eggs, whatever suits your style. Try to snag some local personality to be present to help you celebrate. Have a long distance sailor discuss cruising from your homeport. Perhaps a local TV weatherman can give a hurricane seminar on the dock or help a local sport shop put on a swimsuit competition or fashion show….keep the 'buzz' going.

Any stories about your boat or what happens aboard can be news...and interesting to somebody. I have one story about a special wedding aboard that I often tell reporters looking for a human-interest angle.

It was a first wedding for each partner, a nice couple in their late 30's. After the ceremony, everyone was standing around on the foredeck congratulating the newly wedded couple. I was standing by the ships bell, dressed in my starched whites.

When the congratulations slowed with an awkward pause, to liven things up I said to the wedding party, "On ships we ring the bell to signify a "watch change" and this is certainly one of those."

Whereupon, I rang the bell eight times, ding-a-ding, ding-a-ding, ding-a-ding, ding-a-ding and then watched horrified as the bride broke into hysterical sobbing. What had I done?

"No, no," she cried, "It's alright, It's OK," she said still crying. Then slowly regaining her composure she went on.

"Two years ago, before my brother died," he said, "Sis, when you get married I'll ring a bell for you!"

Then we all cried….

Chapter 7

Insurance, Safety, & Pitfalls

"In certain trying circumstances, urgent circumstances, desperate circumstances, profanity furnishes a relief denied even to prayer."
Mark Twain

Insurance

One of your biggest expenses will be insurance. Shop around; in fact you may have to, just to find a company who will write a policy for you. What, you say? Boat US writes my insurance now. Yes, but they do not write "commercial or charter boat" insurance.

If you are planning to use your pleasure boat for commercial charter, <u>you will need</u> a new insurance company.

Some companies will allow a limited amount of occasional charter time for pleasure vessels, but as soon as they learn you are in the business of chartering, you will need a new insurance company.

Hull, Property, and Liability Insurance

Your charter operation needs two types of insurance, Hull and P&I. Hull insurance covers losses connected with the boat itself, her hull, machinery, accidental contact and damage to docks, other vessels and loss of the vessel whether caused by accident, charterers and/or workers.

P&I is Protection & Indemnity from liability as the consequence of a loss, damage or accident. Your actual policy will spell out the special conditions, deductibles and premiums. If you hire crew to work aboard, whether it is captain, mate or stewardess, you will need crew coverage added to your insurance. Crew coverage is not cheap.

Your cost for hull insurance will usually be many times more than the cost for P&I. For example, you may pay a $2,800 premium for hull insurance on a $1,000,000 vessel and $1,700 premium for P&I for a million dollar policy on the same boat including the mate. It is all a matter of calculated risk to the company.

There are thousands of agents representing hundreds of companies writing insurance policies. Few of them however write commercial insurance policies for charter boats.

Fewer still write policies for charter boats over ten years old and still fewer write policies for boats with other than fiberglass hulls (wood, steel or aluminum).

All agents that do write charter boat insurance deal with an underwriter for the actual policy, no matter whose name is on the agency door or the cover letter you get on the proposal.

Recently there have been only five major underwriters, writing yacht insurance.

ACE (used to be CIGNA)
Markel (Markel American Insurance Co.)
INA (Insurance Company of North America).
Great American Insurance Co.
Chartis Insurance Co.

In the good times, many companies jump into the marine insurance pool to earn premiums, which they then invest in the stock market or other investments. That's right; they use your insurance premiums to invest in stocks. Companies cannot just sit on the premium money waiting for a loss. Your premiums must earn extra money, which is available to help pay the claims and dividends to stockholders.

In bad financial markets, companies cut back writing marine insurance because they cannot justify their payouts from the premiums alone. It then comes down to the heavyweights (those five above) who are in the business come hell or high water, and who structure their premiums to carry on the business during hard times.

Companies under the umbrella of Lloyds of London also write marine insurance policies, but all their policies are through a broker (middleman) and they are really an insurer of last resort, and their premiums will be just that, premium.

Many agents offer insurance, but ultimately all the policies go back to one of the underwriters mentioned above.

How do you find an agent representing a company that writes charter boat coverage? Pick up the phone and try your present agent first, excepting Boat US, Allstate or USAA. Ask your agent for a recommendation. Then look in trade magazines covering the charter boat industry for agents writing commercial insurance. My agent is C.J. Mahlstedt at Island Wide Marine Agency Inc. in Ormond Beach, FL at

800-559-6651. Maybe he'll give me a discount if I send him your business, call him. (*Fat Chance, the Editor*)

Definitions to Know Before You Shop For Insurance

- Market Value, the price the boat would bring if it were sold today.

- Replacement Value, the total cost to build the boat today.

- Insurable Value, the amount of risk the underwriter is willing to take, usually near the market value of the boat.

In setting the insurable value, underwriters often refer to the "moral hazard," when there is a big difference between the market value and the replacement value.

If a boats market value is $50,000 and the replacement value is $150,000, the moral hazard is $100,000 and underwriters likely will not insure much above $50,000 in case of an "accident."

Surveys

Most underwriters consider these points.

- The age of the vessel
- The reputation of the builder
- The hull material
- Where used, inland or offshore waters
- Where moored
- Diesel or gas power
- An acceptable out of water survey
- The market value of the boat

When You Ask For a Survey

- S.A.M.S, A.B.Y.C or I.A.M.L. are Professional Survey Associations, your surveyor should belong to at least one.

- Have a good clear broadside photo of the vessel.

- Have a short concise summary of your experience written down.

- Have a copy of your Masters license at hand.

- Have money in the bank. Your first year's insurance premium will probably be required all up front.

Lastly, never file a claim unless you absolutely have to. Companies talk to each other and if your risk factor goes up with one company, the others will know and your future insurability will be at risk. Filing a claim will usually result in increased premium next year. I am fortunate in never having filed a claim for damage or a loss. It is one of my marketing tools.

"Never an accident or a claim," is a statement I can honestly use when searching for insurance and it pays off. Protecting your claim history is like protecting your credit score, it equates to money in the bank at premium time. I am very careful and always have been operating my vessels. I've also been very fortunate, you could even say lucky, so far. Accidents do happen, even to careful captains.

Sometimes though, not all insurance claims are bad. Returning from a charter in the Bahamas, a safe and careful captain's boat was struck by lightning in Florida's Hillsborough Inlet. He lost all the electronics and electrical power on an eight-year-old, 60' Gulfstar, (which at the time of the strike was up for sale).

When the captain called the owner with the, "bad news," the glee in his voice gave away the good news, "No one was hurt, and all the electronics were fried and will have to be replaced—by the insurance!"

It turned into a $40,000 upgrade courtesy of the insurance company. Not terribly bad news for an eight-year-old boat on the market with old electronics.

Liability Insurance

Why bother with liability insurance? Why not self-insure? If you own the boat and that is all you can lose, maybe you can stand the loss. But if a judgment goes against you, they could take your future earnings as well as your boat. You could be poor the rest of your life. Get liability insurance.

Regardless, you will have to purchase P&I insurance to operate a commercial business from commercial or municipal docks. You must be insured to protect the dock, marina or owner. Your dockage agreement will spell it out, sorry.

P.S. Never let your boat sink due to your own negligence. Your insurance may not pay off. There is an old insurance legend that may or may not be true, but why push your luck.

While an owner-captain was gone, a broken dock hose aboard his boat flooded it with fresh water and it sank. The boat was fully insured but when the adjuster showed up he tasted the water in the bilge, and since it was fresh, he denied the claim. Turn off your water hose when you leave the boat....

Safety Issues

If you are operating a boat carrying six passengers or less the Coast Guard probably will not inspect your vessel, even if you ask. They simply do not have the time and budget to perform these courtesy safety inspections anymore.

The CG Auxiliary will inspect you however, and you should have a fresh safety inspection every year. These are normally free; just contact your nearest Coast Guard Auxiliary, U.S. Power Squadron, or other boating organization.

Securing an annual inspection works for your own protection. First it alerts you to any deficiency in your safety issues; second, it can help you manage any maritime liability should you ever land in court following an accident or other problem. Having a current inspection certificate showing that you always carry the required safety equipment and have it re-certified each year shows your professionalism and concern for the safety of all those aboard, guests and crew alike.

You should know that as an owner of a vessel, that you are personally responsible for the seaworthiness of your vessel anytime you have crew or guests aboard. Maritime law decrees that if you provide a vessel; it must be fit and seaworthy for its voyage even if it is only around the marina. You and the vessel may be liable in case of injury or death of a crew, guest or passenger. Anytime you hire crew for your boat you are liable for injury or death while they are working on your vessel and suffer negligence on your part. You will pay an extra premium for insurance coverage for your crew so be prepared.

You will of course need at least the minimum required safety equipment aboard your vessel as determined by current Coast Guard regulations for your size of vessel and number of passengers carried.

Life Vests (personal flotation devices or PFD's) should be stowed in overhead racks if possible, easily accessible by passengers and crew. Your commercial boat must have commercial quality PFD's, not what you buy at Wal-Mart, on sale. The PFD's must be visible, handy to get at and yet protected from dirt and spray. Wash occasionally to keep them looking new.

Don't forget one for yourself and your crew. Exceed the minimum requirements by having extra PFD's sized for children, and have a couple of approved throwable life rings even if not required on your vessel and add additional fire extinguishers where it seems appropriate. Make sure your flares and fire extinguishers are within their current valid life.

Get a local fire extinguisher company to check and certify your extinguishers annually. Dispose of out of date flares legally (take to Boat US for disposal), do not dump them in the trash.

If you have radar aboard, always turn it on and have it operating anytime you leave the dock. Even on a clear day it could give you advance warning of something amiss ahead or astern. Moreover, having a safety device (radar) aboard and not using it could well prejudice the case against you should you ever land in court. So, when you leave the dock, turn on your radar and log it.

Pitfalls

Whenever your vessel is operating, always be on the alert for other boaters not paying attention. On any busy weekend I see boat operators distracted by talking to their passengers, talking on a cell phone or radio, fishing while underway, relaxing at the wheel and general inattention.

Some small boat operators have no conception of the speed and momentum of a larger vessel and that many times their small boat cannot be seen by you over the bow of your larger vessel.

A friend on his very first professional job heading up the ICW had a small boat pass close aboard his port side. The boat then turned quickly under his bow. His big boat's bow caught the smaller boat by the stern, plowing over it and throwing its operator in the water. No one aboard my friend's boat saw the accident as it all happened under the bow of the larger boat. They hardly felt it on the yacht.

Fortunately the small boat operator was only wet and shaken up, not injured. By the time the authorities completed investigating the accident the operator readily admitted to not paying attention when he cut in front of my friend. Eventually the big yacht was cleared, but everyone lost time, money and reputation, all from a moment of inattention.

Then there are boat operators who mistakenly think that they have the right-of-way when actually they are the burdened vessel, as in overtaking situations. Sometimes sailboats, even if they are under power with no sail up, will jeopardize their boat and mine by powering directly down the middle of a narrow channel. As a sailor, I am sometimes very ashamed of my brethren.

However, I am more often outraged at the lack of common sense of some motor boat operators, particularly rich one's, driving new boats doing the same thing and having no regard for the rules of the road or their own wake.

There seems to be a prevalent lack of common sense afloat around the time of holidays and full moons, watch out!

* * * * *

The islands to windward (East) of St. Thomas, are the Lesser Antilles or Windward Islands. When going there we speak of sailing "down island," i.e. sailing down the chart to windward. The Leeward (lü ard) Islands are the Greater Antilles consisting of Puerto Rico, Santo Domingo, Jamaica and Cuba, all to leeward (West) of St. Thomas.

Chapter 8

Expenses

"I'm Captain Jinks of the Horse Marines, I often live beyond my means; I sport young ladies in their teens, to cut a swell in the Army."   William Lindgard

Start Up Expenses

You need signs for your boat, signs for the dock, and perhaps others. Most of these are one-time expenses, at least for the first year or so. Shop sign shops for the best deals. There is a lot of variation in how different shops work and signs are priced accordingly.

For full color signs, find a shop that services real estate firms. These shops will be easier to work with and usually be a lot cheaper than "signs in a day" shops. Any real estate office should be able to give you a recommendation.

Practically all signs today are computer generated, producing stick-on letters or film that can be applied to many surfaces, even your vessel.

The advantage over the old sign painter working by hand is you can get multiple copies of your signage for just the cost of material. Ask the sign maker to keep your design data so you can duplicate the signs later if you need more copies.

Rack Cards

Start with at least 5-10 thousand copies. Shop around. Go on the Internet and look for printers, rack cards. Your goal is to spread these rack cards around without wincing about the cost every time you hand one out. Hand a rack card out every time you would normally give out a business card. It costs more but tells so much more about your business than a business card and is harder for potential clients to lose.

Once you have an established business name with a website you will be contacted by others selling paid listings on their web sites. Most paid listings are worthless to you. Your best results will come from your own website, which contains the key words, phrases and tags that search engines look for. (Hint, your 20-word description will contain most if not all of your key words for search engines)

Find enough money for the appropriate advertising, and put it where you get direct contact with your customers. The Internet, rack cards, flyers, news stories, and personal contact.

Most magazine ads, general tourist brochures and the like do not have the pulling power that a rack card or brochure directly in a prospects hand has. Try to find ways to make that direct connection.

Other Expenses

You will have to plan for dock rent, deposits, licenses, Chamber of Commerce dues, boat maintenance, haul-out costs, painting, new equipment, insurance, fuel, operating capital, memberships, uniforms, ships stores…the list goes on. Secure enough financing before you start to see you through the first few months.

Very few businesses make enough starting out to pay all their bills. Don't let yourself fail simply from lack of planning your costs up front.

IRS Schedule C

Use a government financial model to set up your business books. Don't panic. I'm only suggesting that you use the IRS Schedule C expense categories as a guide in setting up Microsoft Money, Quicken, or any other system you use for your accounting. Even if your accounting system is a spiral notebook, using the Schedule C headings will save you time and grief at tax time and give you a handy guide for monitoring your business expenses.

A personal recommendation here: I enter all my expenses every day after work. Five minutes at the computer or notebook entering any money you spent that day and a listing of any money you take in takes care of your bookwork. You can unload your pockets of receipts and after entering them in your system just stuff them in an envelope labeled with the month and year. Then monthly or quarterly as needed, you have everything neatly categorized and available on your computer or notebook.

A copy of the IRS Schedule C is in the Appendix, and following are Schedule C categories of expense and how to handle each of them. The numbers of the headings refer to the numbers on the IRS Schedule C.

Expense Categories
8 Advertising

If you expect to take in $60,000 a year, spend at least $6,000 on advertising, rack cards, business cards, uniforms, etc. more if you can afford it. <u>Your business will make no money if you cannot get people onboard your boat.</u>

9 Bad Debts

Avoid Bad Debts by taking only credit cards or cash. Take

checks only from someone you know and can track down if the check bounces. If you make your payment options clear from your advertising, you will have few problems. Credit cards are your best insurance against bad debts.

10 Car and Truck Expenses

Your car or truck expenses are a deductible expense when it is engaged in servicing your business. Transportation from your home to boat is not a business expense. Transporting crew is allowable.

You cannot deduct the cost of buying the vehicle, but you can depreciate it out and if you decide to lease, some of that cost is deductible. Keep good records. Note the mileage on January 1st (or when you start your business), ending with mileage on December 31. Keep gas, oil, maintenance receipts, repairs and misc.

11 Commissions & Fees

The fees or commission you pay brokers to find clients, (usually in the 10-20% range) is deductible. The broker will collect all the money, keep his fee and send you the balance. Keep track of what you are owed. Brokers sometimes play fast and loose with your money.

Find a friendly hotel clerk who can send you business. Pay them for each charter or, better yet, from time to time offer a short afternoon cruise for several of them and their families. It's a commission, not a bribe, they will have a better idea of what you do and your cost to provide the cruise is deductible.

12 Depletion

This refers to stocks of oil or other natural resource that is used up by your business. You do not have a depletion expense.

13 Depreciation

You can depreciate your vessel, deducting the cost of her electronic equipment, running rigging, fishing equipment, your dinghy, truck, or other vehicle over it's useful life. Depreciation can be a trap that costs you when you sell the business, assuming that you eventually sell.

In the beginning, before you are making big money, you probably will not need the depreciation deduction. Get professional help if you want to depreciate any equipment you own.

14 Employee Benefits

If you have no employees, no benefits. If you have employees, get expert tax advice.

Do not, repeat, Do Not Fail to send in employee withholding taxes if you have employees. Money withheld from employee checks for taxes and social security is not your money. Protect yourself by opening a tax account at your bank and make deposits to it each time you deduct withholding taxes from your employee's wages. You will then have the money available each quarter when you must pay those taxes to the government.

15 Insurance

Insurance is one of your major business costs. List here your vessel insurance costs, Hull, Protection and Indemnity, dock liability, crew liability, vehicle insurance, loss of business insurance, uninsured boaters insurance, and business life insurance on you if you are critical to the success of the business.

16 Interest

If you maintain a mortgage on your vessel or purchase anything on credit where interest is charged, you can write off the cost of interest. Interest on business loans, equipment loans, Banking fees, credit card fees, things like that are deductible. Late fees are not. Banks give you a summary of interest paid at the end of the year statement. Use that number, it has already been reported to the IRS.

17 Legal, Professional Service Fees

Any fees paid to an attorney, CPA, financial consultant, web page designer, business consultant, tax advisor is a deductible expense. The cost of this book is a deductible expense for your business.

18 Office Expense

If you maintain a shore-side office you can deduct the rent, electric, telephone, Internet, and cleaning costs here. Home office expenses are listed under line 30. The rules are changing on home offices and you might be better off not trying to squeeze out a few more dollars in deductions without professional help.

19 Pension

If you contribute to an employee 401(k) profit sharing or other pension plan, list those expenses here.

20 Rent or Lease

<u>Dock rent or lease payments</u> for space to dock your vessel or park your business vehicle are fully deductible as long as you are not living on your boat. When your boat is also your home only a percentage of the dock rent is deductible. A tax advisor could advise you on this. However, if you are <u>required</u> to be aboard 24 hours, sleeping space would not be "living aboard."

21 Repairs and Maintenance

Keep good records of time and materials spent repairing and maintaining your vessel. This is not the area to deduct the cost of a new fish finder if you did not have one before. In this category you list your costs for paint, cordage, spare parts for things that break and wear out, oil and fuel filters, technician fees for servicing your mechanicals or electronics, fire extinguishers, engine overhauls and tune-ups.

22 Supplies

Fuel, bait, ice, soft drinks, snacks, food if you are serving meals, toilet paper, soap, cleaning supplies, paper products, brochures, your rack cards, and gift items for resale. Other things, even flowers are deductible, if you routinely have them aboard.

23 Taxes, Licenses

Sales taxes on your charter fees, departure tax, city, county or state occupational licenses, Coast Guard license fees, car or truck license fees for vehicles used solely in the business, all fit this category. If you are 62 or older some jurisdictions waive business license fees for seniors, ask.

24 Travel

Any travel, (fly, drive, boat) to another location in connection with your business is usually deductible. Record the fuel, labor and other costs and enter them here. If you go out without clients one day to scope out a new fishing spot or seek out a hidden anchorage to use, the cost of your fuel used, crew wages, and other expenses are deductible and should be listed here. Traveling somewhere to look for a new boat to purchase is a legitimate business expense, as is taking your vessel somewhere else (deadheading) for repairs. Deadheading costs can include wages, labor, fuel, fees, and <u>lost revenue</u>.

25 Utilities

Electricity, water, trash removal, telephone, cell phone, internet services are all deductible expenses.

26 Wages

If you have crew or pay wages to anyone you will deduct the payroll cost.

If you have employees, find out all the rules in your area. It is very important not to get crossways with the federal and state governments over employee wages and benefits.

Usually the local Chamber of Commerce or any other business owner can send you to the right offices to learn all the requirements for your business. Be straight with them, ask questions and get answers you understand. Leave no room for error. Errors here are expensive.

27 Other

List here any expenses that don't apply to any of the previous categories. Some examples are your time and expenses to train a new captain, crewmember, or pay for equipment lost overboard by clumsy guests.

Chapter 9

Making It All Pay

*"There is a fine line between fishing
and standing on the shore like an idiot."*   Stephen Wright

Income

How to Figure Your Prices to Beat Your Competition

Your income – is never enough. It's determined by how well you sell your business to clients and charter brokers. A lot also depends on how well you manage what business you get and how you manage expenses.

If you plan to operate your business from leads you generate yourself, then you will work harder getting clients for the business than operating the business itself. Great fortunes are made by multiplying the power of many people working toward a common goal.

An individual working alone to create a single product or service rarely makes a fortune.

If you have many brokers working for your business along with past clients spreading good recommendations about you and if your own advertising efforts are effective, then you will find all the business you need. If you try to do it all alone, you may go many years before becoming an "overnight success." In many ways, your success will depend upon how well you relate to people, brokers, charter guests and prospects.

Your Daily Costs

You need to know your break-even point. To find it, divide your estimated yearly expenses including your salary (what you take out of the business) by the number of days you expect to operate the business. To make your estimates, talk to your insurance agent, dockmaster and mentor then calculate your fuel and maintenance expenses and crew expense.

For example; In the APPENDIX is a sample copy of IRS's Schedule C. I've put in some sample figures to arrive at a hypothetical daily expense cost for a small charter business. It shows that to pay me $40,000 a year salary I must generate $221.25 per every working day in revenue.

Divide the annual budget of our hypothetical business ($69,030.00) including your salary by 365 days if your business is open every day, 312 days if you are open six days a week, 104 if you open only weekends, etc. From that figure, you can determine your minimum sales per day to stay in business. This is very important to know.

Remember, that for every day you fail to book a charter or take in some money for your business, that days' income is lost forever. Your boat is like a hotel room, you cannot make up for empty nights. Obviously, if your competitor in the next slip is charging less and going out more you may have to meet his price or find some other advantage your vessel has over his and promote that advantage, in order to have the necessary daily business to meet or exceed your break-even point.

If it comes down to price only, the key is knowing in

advance your fixed costs so you can decide whether to leave the dock for less money to get the business and yet, not burn up all your capital just meeting a competitor's price.

If he goes out of business because he is not covering his fixed expenses and you are, your business likely will succeed. Nevertheless, the goal here is not to drive your competitors out of business. The goal is to grow your business by finding new customers your competitors do not know about. Reach out!

Credit Card Acceptance

This is critical. You must accept credit cards or you'll be turning away business every day. *Every phone call to your business to book a reservation should end with you taking a credit card number.* It creates a moral and financial contract between you and the caller. They are much more likely to show up and honor their reservation if they've given you a credit card number with their reservation. Without it, if they change their minds, you lose. It's that simple.

I currently use my regular bank to process credit cards because I get all my financial information on one statement. It's as simple as phoning an 800 number, following a verbal menu and punching in the numbers of the card, the expiration and the amount.

However, when I first started, my bank wanted too much money in fees for my small operation. I found another company where I could just phone in my credit card charges. Ask around other small merchants in your area. Servicing companies all charge about the same to handle your credit card business, but watch out for annual fees and monthly fixed fees for handling your accounts. Some companies and some banks maintenance fees are not reasonable for a small business.

I accept Visa and MasterCard, nearly everyone has one of those and the fees are less than American Express or Discover.

Currently I pay, on average, a little below three percent, up or down depending on my volume for the month. Add that cost to your prices. One advantage to accepting credit cards is; one does not have to make daily trips to the bank to deposit cash or checks. It's all done over the phone and you learn immediately if the card is good and valid. That alone is worth the three percent to me. You are also less likely to be robbed or leave your deposits on the bar stool after work.

Accounting Systems

I use Microsoft Money on my computer. Quicken or QuickBooks work just as well. If you are not using one of these systems, buy a simple accounting system package at your office supply store.

Even a spiral notebook will work if it is set up properly. Set up your chart of accounts to suit your business, keeping in mind the categories on Schedule C. Group your accounts so you can pull a total for each Schedule C category. It will make life a lot easier at tax time.

Earning Tips

Your tips and those of your crew depend on you. Unless you have worked at a job before where tips were given, the first time or two that someone lays twenty, fifty, or a hundred bucks in your hand you get quite a kick from it. I was past 40 when I began chartering and never in my life up to then had I ever received a tip for working.

My very first charter in the Virgin Islands was two couples who came aboard for ten days over the Christmas holidays. They were younger than I, rich, and very difficult to please. I think they had a silver spoons permanently implanted in their teeth at birth. We struggled through the first few days, me trying to cope with difficult people on my first charter and them trying to make the world run according to their rules.

Near the end, we achieved a sort of truce and I think they actually began to enjoy the charter. The boat, the wind, water and beaches all worked their magic.

When their time was up, I was very glad to see them go but was taken completely by surprise when the "most difficult one" pressed a wad of bills in my hand as he stepped off the boat. They were gone before I had a chance to react, which probably would have been to refuse his tip.

I was really pissed at them but imagine my surprise when I found $250 in my hand and the cook got $100, this back when a dollar was worth something.

It changed my attitude completely toward tips. Now I look forward to the challenge of earning tips from difficult customers. The more demanding they are, the more I work on them for a substantial tip.

How is that you say? Well, I smother them in service. Their every wish that I can anticipate is immediately provided. I chat with them about their business and flatter them on making enough to charter our vessel. I try to become their best friend during the trip, making them feel "special" being able to take a trip most people cannot afford.

I contrast that with simple comments during the charter about how little mates and stewardess earn, comparing what they do to a good waiter does in a first class restaurant. (This is what it seems we sometimes are.)

If a good meal deserves a 15 to 20 percent tip, then how could serving someone all day, one on one, be worth any less? This at least gets them thinking about tipping and that is half the battle.

In my experience, people from big cities naturally tip more generously than those from smaller towns. Guests from the Midwest and Europeans sometimes hardly tip at all. Sometimes you just have to adjust your prices if you know in advance there will not be tips offered.

Also it helps if one of the people aboard is treating all his guests, then he usually is a big tipper to further the impression on his guests. But some never tip, it happens.

Some people with no experience chartering boats do not expect to tip the crew or the captain. Many Europeans do not tip assuming it is included in your price. (adjust your prices) Accept that gracefully. Smile and ask them to invite their friends to visit your boat sometime. You want them to leave feeling great about the trip with you so they will talk up their trip to others. Sometimes that is the best tip of all.

Some guests are especially tight and will do anything to avoid giving you a tip. These people really get my motor going. They're a personal challenge.

Once we had such a family aboard in the Virgin Islands over the holidays. (Why are all the problem charters at Christmas?) All week long my mate and I catered to their three kids, we took them snorkeling, hiking, and showed them all manner of things to occupy their time so Mom and Dad could have a real vacation.

When it came time for them to leave, I offered to take them around to a landing in Lindberg Bay just across from the airport terminal. As they were getting in the dinghy for the short trip ashore, I quietly asked my mate if she had received a tip.

"No," She whispered.

I knew what I had to do. Upon reaching the dock, I unloaded the dinghy with their luggage and, seizing his suitcase, I walked them across the road to the terminal.

The wife kept thanking me for bringing them to the terminal, saving them an expensive taxi ride and making their departure so much more exciting.

At the gate, I stood with them while they checked in, firmly holding his bag in my hand, never letting it touch the ground. They were checking in their luggage and the Missus saw me holding the bag and instantly understood why.

Turning to her husband, "Did you give the crew a tip?" she asked?

"N-n-o, I didn't have any change." He stammered, turning away, embarrassment coloring his face red. She quickly dug her checkbook out of her bag and asked me what would be a fair tip.

"Ten to 15 percent of the charter cost," I replied. She wrote us a tip, nearer to 20 percent, thanking me for the wonderful time she and the kids had.

I gave up the bag, thanked them all for coming and smiling, walked away. Looking back, I saw she was giving him hell as they walked to their plane. I'll bet it continued all the way home.

Oh, and the tip I got from the "difficult one" I spent on a waterproof dive watch I still wear, reminding me daily there is no situation so difficult that it can't be turned around.

To boost your straight charter income, look for some other income streams to help carry you over the slow times and flat periods that every business encounters.

Other Income Sources
- Parties on board
- Catering on board
- Bed & Breakfast aboard
- Weddings aboard
- Filming for movies and media commercials
- Funerals, scattering ashes
- Reunions aboard

In the Virgin Islands, the summer and early fall months are the down season. Many crewed charter boats use that time for maintenance, repair, upgrades, and crew vacations. Some move their boats, going north or south to less hurricane prone weather areas. Some go north to Maine or other points where charter opportunities are greater.

In the Northeast and Midwest many captains work the summer season from May to October, then, lay up their boats for the winter. Some migrate to Florida or down island to find a job aboard a boat in warmer waters.

One friend ran his charter boat in the Virgin Islands from December to April, sailed it to Maine in May, and then managed a small hotel there until it closed in October. He then sailed the boat back in time for the boat show in St. Thomas. His boat was an attraction at the hotel in Maine, offering day sails and evening cruises every week. The almost continuous operation was very hard on his boat and crew. However, it didn't seem to faze him. He liked to be busy, worked his crew hard, and was able to pay to have all the extra work done on the vessel to keep it well maintained.

Other captains work on a boat in the Bahamas, Virgin Islands or lower Caribbean during the winter and then summer in Newport or some other New England resort area captaining a day-sail boat.

In Newport, St. Martin and other locations the current rage is day sails on former 12-metre and "J"class racing boats that have been updated and inspected by the Coast Guard. The captains take a boatload (usually twelve tourists) out in the mornings and afternoons and actually race against one another with these wonderful classic old boats. They race for blood, too. No faking it for the tourists.

Sailing Around the World on Charter

Another captain, who grew tired of the usual milk run through the islands every week, dreamed of sailing around the world. He contacted previous charterers, told them of his plans, and invited them to sail with him on different legs of the trip, for money of course.

The plan worked brilliantly until somewhere deep in the Pacific he ran into a typhoon that stressed his Columbia sloop so much the hull began to separate, and then it split from the bow downward toward the keel.

He and his charter crew on that leg were barely able to stuff the opening enough to keep the water out long enough to reach a port to have the damage repaired. Their charter turned into real sea story to take back home.

Following the repairs the skipper continued the rest of the circumnavigation arriving back in the Virgin Islands three years to the day after leaving.

The plan you work out just may include doing something a little off the edge. But if you enjoy it and can make it pay, why not?

Bed & Breakfast Aboard

In Florida, December is a flat month for chartering except for Christmas and New Year's. Everyone is busy with family and shopping. I was sitting on the aft deck one early December day, reading my newspaper killing time. I was so bored even the classifieds held my attention, and then I saw a heading for "Romantic Retreats." Bingo! That little light bulb lit over my head. That exactly describes our yacht, it's a Romantic Retreat!

Our classic boat has two beautiful staterooms, private heads with showers, air-conditioned and a TV-VCR in each space. Hurriedly, I put a very small ad in the newspaper classifieds.

Bed & Breakfast on a classic yacht.
Downtown Ft. Myers 239-826-2457

The day the ad came out, I received 22 calls. A lot were just curious, and one wanted to know if I supplied the girls. I took names and addresses and sent each caller (except him) a brochure. That day alone I booked six future nights on the boat with credit card numbers to guarantee the reservation, just like a hotel. The ad only ran for three days and by the time it was over, I had every weekend booked for December and the whole boat booked for New Years Eve and the following day for double my usual rate of $125/night for two.

That ad and others similar that followed rapidly evolved into about 40% of my total business. I was able to work the vessel all day on charters and then have people stay aboard at night. The old girl was working 24/7. Suddenly that $40,000 a year was within reach.

After the B&B business really got going we found that by dropping the price to $99/night for two, the increase in business more than made up for the loss from the original price. On holidays and special times, we increased the rate with no price opposition.

I was not living aboard at the time, so effectively guests had the whole boat to themselves. I never book the staterooms to two couples who did not know each other, i.e. strangers. I felt they would be uncomfortable with that arrangement; however, it might work on a larger boat.

I have heard of other yachts doing the same thing and charging much more. Your yacht and your location will guide you what to charge. If you have the *Honey Fitz*, J.F.K's ex-presidential yacht docked in Washington, DC, who knows what you can charge.

To answer your question, no, I've never had any problems at all with guests tearing up the boat or stealing things. B&B people are respectful and in our case seemed to be honored to be able to spend a night aboard a classic yacht. Actually many guests leave the boat cleaner than they found it; with the galley shining, beds stripped and heads wiped down. On all day trips, guests often make crewing aboard part of their experience, wanting to be a part of the ships company, if only for a while. To further that experience I had heavy white Navy coffee mugs made with the ship's name for sale. Nearly every time I served coffee in those mugs I sold some....

You must provide a convenient, safe and comfortable experience if you decide to go the B&B route. Heads have to be antiseptically clean, smell clean and work easily, no dinky pump handles here. I have Galley-Maid® heads and they look and flush like a regular home toilet except they flush electrically with a button driving a remote macerator pump.

Of course you flush into a holding tank. Many guests will ask. Using fresh water to flush will help to eliminate odor and staining of the fixtures. Use your on-board fresh water or a shore hose for fresh water at the dock.

Your bathrooms (heads) aboard must be squeaky clean and smell clean with plenty of quality towels and tissues conveniently at hand. Comfortable roomy bunks are required. Some will put up with narrow Navy type berths, but for the big money and repeat business you need comfortable bunks or beds and quality linens. Romantic activity is going to happen onboard, make it special for them and they will be repeat customers.

Showers must be large enough for an average person to use. Separate shower stalls are preferred over the teak grating in the floor where everything gets wet. Anywhere it's warm, air conditioning that works quietly and very well is necessary along with shade or a covered, nice place to sit on deck.

Your ship's décor is important. Simple, décor without clutter works best. Have nice coverlets on the bunks with plenty of clean, fluffy pillows. Clean, new towels folded "just right" in the head and a vase of fresh flowers will help combat any boat smell. But, if your boat smells, clean it up. Soap, water and elbow grease works. Then ask someone else to give it the smell test, you may not recognize your boats smell anymore.

Make the B&B experience aboard one that your guests will rave about and want to repeat. I have one local couple who book a night aboard nearly every month as their night out, away from the kids. To them it's 'their special romantic retreat.'

We provide coffee, juice, milk, bakery sweet rolls, bagels, fresh fruit and jam in the galley. Guests help themselves to the rolls and fruit and turn on the coffee pot when they are ready. I buy everything at the supermarket the day before, discard the packaging and arrange everything in the galley before leaving for the night.

If you are so inclined and your boat is suitable, you could whip up a regular gourmet breakfast aboard and increase your rates.

Generally, I check people in around four or five in the afternoon, just after coming in from a cruise. They are welcome to stay until noon if we do not have a trip scheduled for the next morning. If we do have a trip scheduled, I can usually get them to buy passage for the day trip, doubling my income for the day. Otherwise, they have to be off the boat by nine a.m. There are very few problems from the timing. As long as they know and understand in advance what we are doing, most guests are very cooperative.

We've had dinner parties aboard, never leaving the dock. I contract with a caterer who supplies all the food and service. All we do is provide the boat, standing by in case anything needs attention. This works for business lunches too, but for those we usually leave the dock for a short cruise, getting the clients away from shore-side distractions and cell phones. If requested, I collect cell phones, telling them it interferes with our navigation (chuckle, chuckle), and the lunch parties have a grand time, being out of touch, and cruising on a yacht. For this we charge a flat hourly rate plus the cost of the catering.

Weddings are generally easy to hold aboard, but it is difficult to charge what they are worth because the happy couple is already paying for flowers, the minister, music, arrangements, limo, travel, hotels and receptions. Paying for a yacht, when the beach is free makes yacht weddings pricey. However, some like the yacht-wedding concept regardless of the cost, so we do weddings quite often. Weddings are usually priced by the hour, with a three hour minimum.

I always note the exact GPS position and Greenwich Mean Time and give the newly married couple a certificate noting exactly where in the world and when they were married.

We have used the yacht as a 'centerpiece' for a dock party at the marina, or on occasion, at a private residential dock. In those cases, the party giver simply wanted to provide something special for his guests, usually for a charity fundraiser. All we have to do is show up, smile, and give short tours of the yacht…. sweet! A few regular charters resulted from those showings as a bonus.

If you have a sailboat, you might offer sailing instructions, familiarization trips aboard and even racing in club events. One boat I know had a slip fronting a major hotel and from only a sign propped up on the foredeck, ran a successful small day charter business long enough to fill his cruising kitty for a trip around the Caribbean.

Moviemakers filming "Just Cause," starring Sean Connery and Lawrence Fishburne needed a sailboat with a very tall mast to provide the reason for a bridge opening in a Ft. Lauderdale chase sequence.

My son was captain of just such a boat and he spent a couple of days rehearsing the shot (at four thousand a day) only to appear for about five seconds in the movie.

Incidentally, that charter came from a local charter yacht broker, another reason to be known by your area brokers.

Many times ad agencies will want a boat for several hours, possibly over several days just to get the right effect for a shot or two. You can watch another of my son's commercial movie charters for CHANEL on YouTube at:

www.youtube.comwatchv=CqF2zm1gVTM&eature=related

Or you might get lucky enough to go to the Bahamas to shoot the annual swimsuit issue for a major sports magazine. Just don't forget to collect your charter money before you sail home.

Funerals

We have never conducted an actual funeral aboard, maybe mine will be the first. However, memorials and scattering ashes at sea is a moving experience and one a charter boat can easily handle. Just be sure to turn the boat into the wind so ashes blow downwind, away from the boat. More than once Uncle Henry's ashes have blown into cabins with open ports after someone got overanxious and failed to wait until they got the "All clear, prepare to scatter."

Federal rules for scattering ashes require that you be outside the three-mile limit from any land. However, often as long as you go out of sight of other boaters or shore side residences, nothing is usually said.

Under federal regulations, the captain must file a letter with the regional office of the Environmental Protection Agency within thirty days stating the date, latitude and longitude, and number of cremated remains scattered.

Bear in mind that what starts out as a pleasant trip with loved ones to scatter ashes sometimes turns into a long, sad trip back. Don't go out so far that you cannot get back quickly. You do not want to increase their pain.

Actual burial of a body at sea is complicated. Federal regulations require you to be at least three nautical miles from any land and you must be in at least 100 fathoms of water. In many areas the 100-fathom line is too far out to be practical. You must also insure the remains sink to the bottom rapidly and permanently.

One of my readers, Capt. Gene, operates a business handling ash scatterings. He's at; affordableashesatsea.com Check on Google for sea burials, there are many others.

Old naval funeral traditions are sometimes followed according to the deceased wishes. The following is quoted from "Naval Customs, Traditions and Usage" by Leland Lovette, Cdr. U.S. Navy for direct body burial at sea.

"...the coffin covered with the American flag; with the union (blue field of stars) placed at the head and over the left shoulder." The ship's chaplain or captain reads the burial service at sea. The Episcopal prayer-book service is commonly used. The ritual ends with the very beautiful and time-honored words,

"We therefore commit this body to the deep, to be turned into corruption, looking for the resurrection of the body, when the sea shall give up her dead and the life of the world to come."

Someone, (a funeral service provider) having a state permit to transport and handle bodies must provide the suitable shroud (omitting the last stitch through the nose) and weighting it with enough iron or lead it so it will sink quickly and permanently. The entire process falls under the Ocean Dumping Act. The fine for illegally dumping bodies is currently $50,000. Not something even Tony Soprano would laugh at.

Once when I was at the dinghy dock in St. Thomas, a fellow drove up quickly in a pickup with a box in the back.

"Did you see the Pilot boat here?" he asked, as I tied my painter to the dock stanchion.

"Yep," I answered, "It left just as I was coming in, had a lot of locals on board, dressed to the nines."

"Well, they're going out to sink the wrong body," he said, and drove off....

Chapter 10

Charter Brokers

"To be successful at sea, we must keep things simple."
Pete Culler

A Salesman for your Business

Brokers bridge the distance between boat owners and clients. They are your sales representatives of the charter business, the movers and shakers. The saying, "Nothing moves until someone makes a sale," could not be truer than in the yachting business.

Charter brokers amass a lot of knowledge about different boats, crews, vacation destinations and procedures. They match that up with clients who want to spend time and money on a yacht vacation somewhere. Charter brokers differ from yacht brokers who are primarily engaged in selling yachts.

Yacht brokers sometimes however, arrange charters for prospective buyers to try out a vessel before getting them more seriously involved in a purchase, whether on that vessel or a similar one.

In the U.S., charter brokers fall into two basic categories; those who represent smaller yachts and sailboats, which will probably be the brokers you may use, and brokers representing a few, primarily larger yachts and Mega yachts.

Large yachts are those up to about 125′, similar to a well-found Broward, Burger, Westship, or Azimut. These boats will charter for up to $50,000 a week including fuel and provisioning. Mega yachts, those generally over 125′ long, can cost $80,000 a week or more for eight persons.

Just to whet your appetite, some Mega yacht captains with all the necessary Coast Guard certifications including certification by the new STCW (Standards of Training, Certification and Watch-keeping for Seafarers) and the experience to match can earn up to two hundred fifty thousand dollars a year as Captain, with tips and gratuities adding up to another fifty thousand more. (There are now over 3,000 Mega yachts worldwide with more building every year.)

Boat shows for charter brokers occur in many places during the year to give brokers and yacht owners/captains an opportunity to mix and mingle, show off their vessel, meet the crews, pick up brochures, develop some rapport with captains and generally have a good time.

It's work, to be sure, tramping up and down docks in the hot sun in some exotic location and trying to absorb and retain the important stuff, but for many brokers and captains alike, it's also a time to socialize and 'gam' with friends, perhaps their only opportunity during the year.

Other brokers representing a few luxury yachts get more involved in the day-to-day booking and business of the yacht, often working hand in hand with the captain or owner to successfully market and manage the yacht's charters. These brokers will assist with selecting a theme for the vessel, perhaps finding decorators and shipwrights to execute renovations to a 'tired' boat or one that shows a lot of use.

Most brokers belong to various associations that provide some consistency of approach to boat owners and clients.

Brokers that book the large yachts and most foreign brokers act independently and follow their own, industry standard rules. If you are trying to charter your large yacht you will need both an experienced captain who has been around and can bring your yacht up to "Bristol Condition" and keep it there and a busy, well-known broker to handle your yacht. Nothing serves like experience in these waters.

Pick up a copy of any major yachting magazine and check the display ads for brokers handling boats similar to yours. Call them up and ask how they can help you, get references of other boats they handle. Then talk to other boat owners about their experiences.

Charter yacht Brokers Association (CYBA) has rigid requirements for representing yachts and handling funds deposited, requiring members to follow ethical guidelines and attend charter yacht shows to keep current with the yachts. Many U.S. brokers belong which creates some uniformity and tends to moderate the sometimes turbulent charter seas.

Check the CYBA website (www.cyba.net) for a list of current and future charter boat shows. Most will allow the public to attend for a fee. Try to attend at least one before heading off to one in your own boat. There are many valuable lessons you can learn and most yacht captains and brokers will be willing to give you a hand and offer advice.

Working with a broker usually follows this scenario. You prepare your vessel for a white glove inspection, making sure everything works properly then take her to a charter boat show (sometimes paying a lot of money to the show promoter for the privilege), spend the next few days onboard selling the boat, your crew and yourself to perhaps dozens of brokers and curious dock walkers.

If you are lucky, you will come away from the show with one or two brokers who like what you have and will work with you, selling your charters to their clients. For this, they will collect (deduct) anywhere from 15-20% of the charter for their fee when they book a charter on your boat.

Contracts

The contract with the charterer will spell out who does what, when, where and how. To book, the prospective charterer will deposit half the charter fee, the fuel advance and the provisioning advance with the broker, along with the food and drink preference sheets.

The usual deposit is fifty percent of the charter fee, all of the estimated fuel and provisioning cost advances. The broker will forward the advances and preference sheets to you with your 50% of the charter fee less his total booking fee.

So for example, out of a $10,000 charter fee with $2000 advance for fuel and $2000 provisioning advance, the broker sends you both advances, plus half of the $10,000 charter fee less his 20% commission. Or $2000 + $2000 + $3000 = $7,000 before the charter ever starts. All you have to do is show up at the designated time and place, with the yacht fully fueled and provisioned for the duration of the charter.

At that time, you get the remainder of the charter fee ($5000) either from the client or from the broker who has collected it in advance of your departure.

That is how it is supposed to work. In theory it usually does, more or less, or boats and brokers would stop being in business. But, reality slips in due to delays of checks clearing, lost overnight mail, charterer departure time and embarkation changes.

Or the broker forgets to send the check, the broker is "out of town," unavailable or the check is drawn on a foreign bank and has not cleared and half a million other excuses which you will learn as you go along. Like any owner-operated business, you will spend a lot of your time 'chasing the money.'

I have heard all those excuses, often more wash up prior to departing on charter, and you will too undoubtedly adding your own. Typically, a charter is booked and then for one reason or another no money is transferred as planned. The broker will call you and say,

"Just collect from the charter party and we'll settle up between us when you get back."

The charterer shows up, claims to have paid the broker his 50% deposit and balks at paying for fuel up front and is unwilling to front the provisioning fee in full because,

"We don't eat much and besides we'll eat out a lot on charter."

Then he wants to give you a personal check on a small town bank in Iowa for the rest of the charter fee, and its Saturday on a long weekend. Of course this example is made up, but reality is often not far behind. I've not heard it all yet.

Or, everything works as planned except without warning the charterer brings along two of his daughters best friends upsetting the sleeping arrangements, causing you to run out of all soft drinks three days early, and both girls have special dietary eating habits.

Fussy eaters drive your chef bananas because there is no corner store out on the water to run to for special food requests. My chef once bought up all the frozen chicken in Tortola for a party of eight who did not eat red meat; this after she had stocked up on specialty steaks for grilling.

Or, at every planned stop on your itinerary which the charterer has sent you earlier, he invites aboard half a dozen (or more) of his friends all waiting on the dock as you pull in.

They eat, drink and trash the boat, keeping you and your crew up half the night cleaning up when you should be sleeping to prepare for an early departure to the next party many miles distant.

Or, it's the day before your charter; you and your crew are cleaning, making up staterooms, scrubbing everything in sight squeaky clean in preparation for the charter. The chef is off to the store to provision and guess who comes waddling down the dock, trailed by his brood and a parking lot attendant pushing a cart full of luggage.

"Hi Captain," he says brightly, "Our flight got in early so we decided we'd like to stay aboard tonight instead of in that stuffy old hotel. It won't be a problem; we'll keep out of your way."

Or your charter party announces that they are nudists and have no intention of wearing clothes for the whole week, which is OK with you except you have a hard time getting them off the foredeck when entering port to clear customs or docking at a restaurant for dinner.

This is not to suggest that all charters and brokers are bad news. Brokers have no control over insensitive charterers or what happens after the boat leaves the dock. Most will bend over backwards to help you have a successful charter because they want the repeat business as much as you do.

When you find a broker you like, work with them and listen to their advice. They have a unique position that lets them see what makes some boats do well and other boats not so well.

If yours is a very small operation, you may not need a broker to handle your business; most brokers are not interested in the small charter or fishing boat that generates only enough income to keep the captain happy. The booking fee for an all day fishing charter that might gross $1,500 is not worth their time, and you can hardly afford to give up 15 percent of your business right off the top.

Brokers are not for everyone. Seek out and talk to brokers in your area and at charter boat shows. As you develop your business you will learn a lot and possibly alter your business plans to more closely fit into the local charter market. None of us has all the answers and the answers change with the economy.

Travel Agents

Travel agents are normally useless to the charter business because they operate under a different commission structure and haven't the time to get to know the boats and where they go. Few agents are familiar with more than a few boats and clients do not commonly call a travel agent to book a charter boat as they might have done in times past.

The Internet has permanently altered air travel and vacation planning, rendering many of the services travel agents traditionally provide unnecessary. If you happen to have a friend in the travel business, perhaps they can steer you some business if your work together on a plan, but don't count on it.

I have a travel agent friend who three times booked our boat in the Virgin Islands for a singles cruise, with her along as tour guide. She got a free vacation twice a year and we got an interesting charter, but the charters stopped coming after she found her man and married him… aboard our yacht!

Chapter 11

Owners & Captains

"It's good to have money and the things that money can buy. But it's good too, to check once in a while and make sure you haven't lost things that money can't buy."   George Horace Lorimer

Switch Hats for a Moment:
This Chapter is about Boat Owners, Captains, and Crew

Congratulations Mr. Owner, you have been successful enough in your business or profession to be able to afford a yacht, maybe a nice, really big yacht. You have worked long and hard to build up your business, hiring, firing and working with many talented people who helped you reach your goals.

Now apply those same rules of honesty, fairness and courtesy to selecting your captain. You are putting an employee (not a servant) in charge of an operating business, one you hope and expect to give you a good return, if not in dollars at least in pleasure.

Be it money or satisfaction, you want to be happy with your decision and selection. Your captain wants exactly the same thing. If you treat him (or her) with respect, courtesy and do what you say you will do, I can almost guarantee you will have a good relationship with your captain and crew, and you will see a happy ship when you come aboard.

However, if your standard business practice is to bend the rules, look for any possible business or moral advantage and not trust anyone, please don't hire a captain and micromanage what he or she does.

Let your captain hire the crew. Sure, he is going to hire his friends if he can, but at least he knows them and their abilities. A good captain will know several crewmembers by reputation or people he has worked with in the past. These people know he can rely on them, and they on him.

It should be obvious but I'll say it anyway, **"Don't put your girlfriend, lover, best friend or relative on board as part of the crew."** Everyone will suffer, most of all, you. And don't try to "salvage" someone by giving them a job on your boat, expecting your captain to work miracles, it won't happen.

Don't second-guess and undermine your captain's authority on board with his crew. You may pay their salary, but the crew works for your captain. Do not make side deals with the crew, brokers, marinas or guests without your captain being fully informed.

Running (being responsible for) a boat 24/7 takes a great deal of trust between members of the crew; the captain and an owner who will let them do their job.

I'm not saying turn them loose, but monitor their work the same way you monitor a trusted employee at your place of business. Set goals, discuss problems, provide resources and monitor results, most of all communicate with your captain.

Your captain is responsible for; 1)<u>Your safety</u>, 2)<u>Your guests safety</u>, 3)<u>The crew</u>, and 4)<u>The vessel</u>, in that order. He must have the final word on safety issues. Your insurance company wants it that way also.

Cash Requirements on Board

I knew an owner who entrusted his captain with his million-dollar plus yacht then refused to advance the boat a couple thousand dollars for petty cash. He expected the captain to pay all the little stuff out of his own pocket and then at the end of the month the owner would pour over the list of expenses and check receipts as if they were his very last dollars.

To add insult to this, the credit card he did furnish for ordinary boat expenses and fuel purchases was often maxed out over the credit limit because the owner was using it for "personal" expenses he didn't want his wife to know about.

This captain continually struggled to have enough money available to buy fuel and pay for boat expenses. After a couple years of this, the captain found a better, more understanding owner, and jumped ship. Still some owners wonder why they have trouble finding and keeping a good captain and crew.

Consider some of the things Captains need instant cash (boat money) for:

- Fuel for the dinghy and water toys
- Spare parts needed in a hurry
- Ice for drink coolers
- Purchasing last minute provisions for charters
- Special food or drink for charter guests
- Taxi fare for crew and guests
- Paying car rental fees for guests or crew
- Parking fees

- Cruising permits in foreign ports
- Paying clearance, immigration and health dept. fees
- Postage for bills and receipts sent to owner
- Office supplies used onboard
- Marine store purchases of supplies used aboard
- Buying flowers, birthday presents, special gifts for guests
- Fishing gear, bait, and guide services for guests on charter
- Masks and fins for guests, water toys for kids
- Casual labor costs for extra boat cleaning and bottom scrubbing
- Buying lunch or dinner out for the crew after a particularly hard charter
- <u>Everything else</u> you can think of.

Transportation

In your other business, you probably don't expect your employees to use their personal cars for your company business. You probably don't want the liability of them using their own cars to pick up you or your clients, taking them to dinner, hauling supplies for your business. Your insurance company doesn't want it either.

Either provide your crew with an appropriate vehicle at your homeport or funds for a rental when needed. Remember it takes extra time to get and return a rental car or van. Time they could be working on your boat instead of shuttling cars.

The same reasoning applies to the dinghy or yacht tender. It should be large enough to do the work needed to support your boat. If you anchor out a lot and run to shore, your crew will need a husky dinghy, perhaps even two, with enough power to get you there and back quickly, in comfort and safety.

Think of it as a lifesaving device as well. You may never use it as a lifeboat, except if it's needed to get you or a guest to medical facilities in a hurry or going into some place where your big boat cannot go or taking it there would take too long to arrive. In those cases the dinghy might be a lifesaver, even your lifesaver. Let your captain participate in selecting the right dinghy and all major equipment for your boat. He will know what the yacht needs for the waters you sail.

My first owner provided the boat I was captioning with a blow up Avon with paddles, suitable only for children's play in shallow water. Against his wishes, I immediately bought a 14′ heavy-duty island fishing boat for a dinghy with 15-horse power outboard using money from our first charter. He was pissed at first but later using the boat himself agreed it was a sound investment. It was really a safety issue not a money issue.

Work Schedules

Most crews on charter work at least 12 to 14 hour days, often longer. Using a 44-hour workweek, that equates to them ending their workweek about noon on Thursday, not counting any weekends.

You may be on vacation and looking forward to spending two glorious weeks in the Bahamas or other cruising destination, and your crew wants you to enjoy it. But, and this is a big one, remember your crew is working 24/7 to get you there and take care of you. They need some down time too.

Every chance you have to give them an afternoon ashore, or time off to visit other friends on nearby boats, or just go off and read a book, give it to them. They will love you for it and your happy crew will be the envy of all your boating friends.

Work your vacation schedules to give your crew some holiday time as well. They also have families and friends…a life. It is not fair to expect your crew to be available to you or be on charter every holiday. Everyone has some place where they would like to go on their time off.

If your travels take your vessel offshore or out of your normal cruising grounds, pay the transportation costs for your crew to go home occasionally on vacation.

Yes, your crew is enjoying the fruits of your labor, going places on board a fine yacht, staying in ritzy marinas, dining in fine restaurants with you and your guests. Nevertheless, it is still just work for them. Recognize that, pay them accordingly, and give them free time when you can.

Pay

Pay for yacht work varies of course, but up to about 100 feet of boat you can expect to pay a captain about a thousand dollars a foot per year in salary. The captain for your active eighty-five foot charter boat should earn about eighty-five thousand per year.

When times are tough, that salary may drop some, but you get what you pay for and if you stay on the cheap end of the pay scale, you will forever be looking for, dealing with and losing captains and crew. Your boat will suffer from lack of continuity in crew and adequate maintenance, costing you more in yard bills at refit time.

The reality is; you need the energetic captain even more when times are tough to successfully charter your boat to help with the expenses, much less make some money with it. A successful captain with a good following among charter brokers and past clients can make even an old dog of a charter boat do tricks.

True, the old boat will cost more in maintenance, and the crew needs extra time, and money for dealing with the old dog, but a good captain with experience will earn it all back and is worth every dollar.

If other boats in your class are doing 8-12 weeks a year of charter, your good captain well known to the brokers may often turn in 12-16 weeks or more. Those extra weeks will more than make up for the salary costs.

When the country goes into a recession, no matter how minor, boats on the low end of the charter scale are the first to suffer from lack of business. The economy does not affect the super rich, usually their vacation and holiday plans move on regardless. However, the business owner whose income rises and falls with the economy will back off from discretionary spending during tough financial times. This goes doubly for their boat.

The boat suffers from lack of maintenance and crews are laid off. Sometimes the crew will drift into other lines of work and be gone forever, while others will work at any price but hate it. As an owner, if you expect to stay in the business of enjoying your boat, in good times and bad, find and pay good crew adequately.

Remember also that having you or your guests aboard are just the same as chartering to the crew. You may, in fact, be more demanding than charter guests paying their own way. And charter guests may tip the crew handsomely, you probably won't.

The members of your crew are trusted employees; treat them with the respect they deserve. Appreciate their efforts and their company when you are aboard, tell them how you would like your yacht operated, listen to their ideas, work together to reach your goals, deal with problems fairly and everyone will do their best for you, remembering that poet Robert Louis Stephenson said it best,

"Old and young, we are all on our last cruise."

Chapter 12

Putting it all Together

"The sea's most powerful spell is romance."   H.W. Tillman

Still with me? Good. If you have used this book like a workbook you have:

- Earned your Coast Guard license
- Found the proper boat, yours or another's
- Researched your market
- Financed your dream with enough money
- Kept to a timetable
- Have a three ring binder of your Charter Plan

You are now aboard your vessel (or someone else's) happily settled in the captain's chair enjoying your new role as Charter Boat Captain. Likely though, you are still somewhere along the way towards that goal, that's okay too.

Keep working on the small details and the larger challenges until you have everything handled and under some control.

Dealing with the Coast Guard, your local government offices and regulatory procedures can be a pain in the ass. Don't let it get you down, never get mad. Be courteous, ask questions, ask for help, and satisfy yourself that you understand all their precise requirements before you leave the building. Come prepared, be persistent and follow through. You will prevail if you keep trying. Life happens; but it never lets us be fully in charge.

If this book has simply given you an overview of what is necessary to reach your goal, whatever that may be, well….I hope you have a better idea of what to do and where to go next.

Review the books and other material listed in the Appendix. Take what you need. The world is full of many great ideas and most of them have been written about at one time or another. Many books are available in used bookstores or your local library. Use every resource to help your business. It is much less expensive to buy a book than to make a mistake. Books are cheap, mistakes can kill you. Books by others I can recommend are:

- "Make Money with your Captain's License" 2008 by David G. Brown (no relation)

- "Making Money with Boats"1996 by Fred Edwards

- "No Shoes Allowed" 1996 by Jan de Groot

- "The Cruising Life" 1980 by Ross Norgrove (I was privileged to know Ross and call him a friend. He was one of the pioneers of chartering and the business of today was laid on foundations figured out by Ross and his friends.)

The Appendix has examples used in the business, addresses where you can go for help, sources for brokers, insurance, captain's schools and Coast Guard help. If I can help, e-mail me at:

captconrad@shipyardpress.com.

If you can, go back, talk with your mentors, and 'gam' about what you both have learned. Keep in touch with your boating friends, share the rewards and difficulties, learn from others and teach them what you know. Practice good seamanship on board and demonstrate competence and courtesy on the water.

If you have an opportunity, teach what you know to some young people interested in the sea and boating. Show by example when you can. Let them see you do what they need to know to be safe and secure afloat.

The rest of the voyage is in your hands. I hope that crew named 'Good Luck' is always aboard your vessel. When you get discouraged as I did and you will also, find inspiration from these words of Karen Blixen's, the Danish author of "Out of Africa,".

"The cure of anything is salt water;
sweat, tears, or the sea."

The End, or your Beginning

Thank you for reading this book. I hope it answered some of your questions and gave you a blueprint for your chartering experience. If you found it valuable please recommend it to a friend or write a review and post it on Amazon to help others live their dreams, Thank you.

About the Author;

Conrad Brown earned his 100 ton Coast Guard Masters License in 1979 after living in Puerto Rico and the Virgin Islands. His first Captain's job was operating a powerful 50' Ed Monk designed ocean cruiser for an absentee owner. Like most yachting jobs, he got a lot of experience and some pleasure from this while the owner paid the bills and enjoyed the boat vicariously.

From there he lived aboard and cruised on his own boats; *Halcyon, Capricornus* and *Caribe Belle* in the Caribbean and made several deliveries back to the states as captain or as crew on his and other's boats, sailing *Capricornus* back to Beaver Lake in Arkansas and ran a charter business there. Another great trip was as relief captain on a 90' Phil Rhodes sloop, Virgin Islands to Newport, port tack all the way.

Later he rescued a 60'- 1936 wooden trawler, *Black Tie*, updated it extensively and put her into the river tour business in Florida. After enjoying the boat and business for several years, he sold that boat and business and wrote this book to help you start yours. Hope you enjoyed it.

Changes as of April 15, 2010 to the Coast Guard Licensing

See

http://www.uscg.mil/nmc/cb_capt.asp

Before you can apply for a license you must obtain a

TWIC

Which is a

Transportation Worker Identification Credential.

Go to this site for an application wizard to help you apply.

http://www.uscg.mil/nmc/credentials/original/original.pdf

National Maritime Center has been going through several new changes in regards to issuing Merchant Mariner's Credentials (MMC).NMC is now producing a new credential which looks similar to a United States Passport.All US Mariners will be issued this credential as a replacement of their past MMD.You will have several endorsements depending on your Seatime and tonnage experience.

Once you have your TWIC you may apply in person or by mail to one of the following Regional Coast Guard Examining Center (REC)

U.S. Coast Guard
Marine Safety Office
Regional Examination Center
800 E. Diamond Blvd.
Suite 3-227
Anchorage, AK 99515
Phone: 888-427-5662

U.S. Coast Guard
Regional Examination Center
U.S. Customs House Bldg
40 South Gay Street
Baltimore, MD 21202-4022
Phone 888-427-5662

U.S. Coast Guard
Marine Safety Office
Regional Examination Center
455 Commercial Street
Boston, MA 02109-1045
Phone: 888-427-5662

U.S. Coast Guard
Marine Safety Office
Regional Examination Center
196 Tradd Street
Charleston, SC 29401-1899
Phone: 888-427-5662

U.S. Coast Guard
Marine Safety Office
Regional Examination Center
433 Ala Moana Blvd.
Honolulu, HI 96813-4909
Phone 888-427-5662

U.S. Coast Guard
Regional Examination Center
8876 Gulf Freeway Suite 200
Houston, TX 77017-6595
Phone: 888-427-5662

U.S. Coast Guard
Regional Examination Center
9105 Mendenhall Mall Rd.
Suite 170, Mendenhall Mall
Juneau, AK 99801-8545
Phone: 888-427-5662

U.S. Coast Guard
Marine Safety Office
Regional Examination Center
501 W. Ocean Blvd. Ste 6200
Long Beach, CA 90802
Phone 888-427-5662

All phone numbers are the same, when you call, ask for the REC center nearest
your home or where you expect to be working.

U.S. Coast Guard
Marine Safety Office
Regional Examination Center
200 Jefferson Ave.Suite 1301
Memphis, TN 38103
Phone: 888-427-5662

Regional Examination Center
Claude Pepper Federal Building
51 S.W. 1st Ave., 6th Floor
Miami, FL 33130-1608
Phone: 888-427-5662
Phone: 888-427-5662

U.S. Coast Guard
Marine Safety Office
Regional Examination Center
4250 Highway 22, Ste. F
Mandeville, LA
Phone: 888-427-5662

Regional Examination Center
Battery Park Building
1 South Street
New York, NY 10004-1466
Phone: 888-427-5662

Marine Safety Office
Regional Examination Center
911 NE 11th Ave, Rm 637
Portland, OR 97217-3992
Phone: 888-427-5662

U.S. Coast Guard
Marine Safety Office
Regional Examination Center
915 Second Ave., Room 194
Seattle, WA 98174-1067
Phone: 888-427-5662

U.S. Coast Guard
Regional Examination Center
Oakland Federal Bldg., North
1301 Clay Street,
Room 180N
Oakland, CA 94612-5200
Phone: 888-427-5662

U.S. Coast Guard
Regional Examination Center
420 Madison Ave.Suite 700
Toledo, OH 43604
Phone: 888-427-5662

U.S. Coast Guard
Marine Safety Office
Regional Examination Center
1222 Spruce Street,
St. Louis, MO 63103-2835
Phone 888-427-5662

Below are some tools and resources I have found valuable. Naturally phone numbers and addresses change over time, so you may have to do a little digging to get the current number if it has changed. For this reason web pages are listed if they exist. All numbers and web links were current at time of publication.

- Your local Chamber of Commerce
- Your state tourism agency, mine is **www.flausa.com**
- National Maritime Center (NMC) **http:/www.uscg.mil/nmc/rec_information.asp**
- CG Application Check List: **http://www.uscg.mil/hq/cg5/nmc/announcements/Application_Process_Bulletin.pdf**
- Captain's School: **www.captainschool.com**
- Houston Marine Training Services: **www.houstonmarine.com**
- Sea School: **www.seaschool.com**
- Adams Marine Seminars: **www.adamsmarine.com**
- National Captains Institute: **www.captains.com**
- To find a captains school near you, Google "earn captains license"
- Virgin Islands Charterboat League: **www.vicl.org**
- Crew news: **www.dockwalk.com**
- Uniforms: **www.estore.smallwoods.com**
- Uniforms: **www.superfineusa.com**
- Boating forum, **www.woodenboat.com**
- Maritime Books: **www.shipyardpress.com**

Ten top reasons CG applications are delayed:

1. Applications – If the application is not completed, it will be returned for correction. Three signatures are mandatory: Section III ("Have you ever...?" questions), Section V (consent of National Driver Registry check), and Section VI (application certification). When the "Applying for:" block is left blank or is incomplete, the REC is left to guess what you want.

2. Drug Screen – A drug screen is often rejected because it does not contain the Medical Review Officer's (MRO) signature, it is a photocopy, or a company compliance letter is not written to meet the requirements of the Code of Federal Regulations, Title 46, Part 16, Section 220.

3. Photographs – Merchant Mariner's Documents (MMDs) and STCW certificates cannot be printed without a photograph. Two passport size photos are needed when applying for an MMD or STCW. (see http://www.uscg.mil/nmc/cb_capt.asp)

4. Physical Exam – If the Merchant Marine Personnel Physical Examination/Certification Report is not complete, it will be returned for correction. Particular attention is paid to the "competent", "not competent", and "needs further review" boxes, which are frequently blank. Often the type of color vision exam given in Section IV in not indicated or mariners who wear glasses and/or contacts submit exams without their uncorrected vision listed in Section III.

5. Original Certificates – Photocopies of essential documents, even if notarized, are not accepted. Only original signatures, those documents signed by the issuing authority (e.g., course completion certificates) or official custodian (e.g., birth certificates) are acceptable. Original certificates will be returned when the evaluation is completed and the REC mails the newly issued credentials to the applicant.

6. User Fees – No or incorrect fees are included with the application. Licensing user fees changed April 15, 2010 Current fees are published in the most recent Code of Federal Regulations, Title 46, Part 10, Section 109 and on the web at: http://www.uscg.mil/STCW/l-userfees.htm.

7. Current or Past License, Document, and/or STCW – A mariner who is holding, or has held, a license, MMD, and/or STCW certificate who does not indicate it in the history (Section II of the application) or does not include a copy of their credentials (front and back) with the application package. This especially applies for renewals and mariners with past transactions at other RECs.

8. Sea Service – Missing or conflicting information on the sea service letter (e.g., not including tonnage or horsepower, the position listed does not agree with other documents in the application package, or conflicting waters). Service should be documented with discharges, letters from marine employers, or small boat sea service forms. If a small boat service form is used, it must be certified and signed by the owner or proof of individual ownership is required.

9. Written Statement – If an applicant marks "Yes" in any block of Section III, a written statement is required. Note that all questions beginning with "Have you ever..." include all past convictions, even ones that may have already been disclosed. Simply stating "on file" will not suffice, statements should include the what, when, where, and penalties assessed for each incident, if it has already been disclosed to the REC, and whether there have been any new incidents. The applicant must sign and date the statement.

10. Medical Condition – Additional medical information is required whenever a medical condition is identified on the Merchant Marine Personnel Physical Examination Report.

Expense Form for your Business

The table below shows some typical expenses from a small charter boat operation. Fill in your estimated amounts and see what your business might net at the end of the year. Schedule C is the model for this chart.

All amounts are dollars.	Sample	Yours
8 Advertising	4714	
9 Bad Debts	0	
10 Car, Truck Expenses	816	
11 Commissions & Fees	1600	
12 Depletion	0	
13 Depreciation	0	
14 Employee Benefits	0	
15 Insurance	4788	
16 Interest	547	
17 Legal, Professional	0	
18 Office Expense	580	
19 Pension	0	
20 Rent or lease	5168	
21 Repair, Maintenance	2901	
22 Supplies	634	
23 Taxes, Licenses	865	
24 Travel	0	
25 Utilities	2376	
26 Wages	0	
27 Other	4041	
Total Yearly Expenses	29,030	
Pay yourself	40,000	
Required Gross income (expenses + pay = gross	69,030	
Daily gross receipts (312 days =six days a week)	$221.25	

Expense Form for your Business

1. Write Your Goals

2. Write a 50 Word Description

Write a 20 Word Description

3. Write a One Sentence Description of What You Do

4. Who Are Your Customers

5. How Do You Expect To

Find Your Customers

6. Who is Your Competition

(Get their brochures, prices)

7. How Much Can You

Spend for Marketing

(Break it down by quarters)

8. What Steps to Your

First Day of Charter

(List of Gets)

9. Timetable

10. Find a Mentor

(list possible mentors here)

Notes to Yourself

www.ingramcontent.com/pod-product-compliance
Lightning Source LLC
Chambersburg PA
CBHW060023210326
41520CB00009B/981